MODELS FOR CONCURRENCY

ALGEBRA, LOGIC AND APPLICATIONS

A series edited by

R. Göbel
Universität Gesamthochschule, Essen, Germany
A Macintyre
The Mathematical Institute, University of Oxford, UK

Volume 1
Linear Algebra and Geometry
A. I. Kostrikin and YU. I. Manin

Volume 2
Model Theoretic Algebra: With Particular Emphasis on Fields, Rings, Modules
Christian U. Jensen and Helmut Lenzing

Volume 3
Foundations of Module and Ring Theory: A Handbook for Study and Research
Robert Wisbauer

Volume 4
Linear Representations of Partially Ordered Sets and Vector Space Categories
Daniel Simson

Volume 5
Semantics of Programming Languages and Model Theory
Manfred Droste and Yuri Gurevich

Volume 6
Exercises in Algebra: A Collection of Exercises in Algebra, Linear Algebra and Geometry
Edited by A. I. Kostrikin

Volume 7
Bilinear Algebra: An Introduction to the Algebraic Theory of Quadratic Forms
Kazimierz Szymiczek

Volume 8
Multilinear Algebra
Russell Merris

Volume 9
Advances in Algebra and Model Theory
Edited by Manfred Droste and Rüdiger Göbel

Volume 10
Classifications of Abelian Groups and Pontrjagin Duality
Peter Loth

Volume 11
Models for Concurrency
Uri Abraham

This book is part of a series. The publisher will accept continuation orders which may be cancelled at any time and which provide for automatic billing and shipping of each title in the series upon publication. Please write for details.

MODELS FOR CONCURRENCY

Uri Abraham
Department of Mathematics and Computer Science
Ben Gurion University, Be'er Sheva, Israel

CRC Press
Taylor & Francis Group
Boca Raton London New York

CRC Press is an imprint of the
Taylor & Francis Group, an **informa** business

Amsteldijk 166
1st Floor
1079 LH Amsterdam
The Netherlands

British Library Cataloguing in Publication Data
A catalogue record for this book is available
from the British Library.
ISBN 90-5699-199-X

Contents

Preface   ix

PART 1
Semantics of distributed protocols

1. **Elements of model theory**   **3**

   1. **Structures**   **3**
      1.1. Examples of multi-sorted structures   5

   2. **First-order languages and satisfaction**   **8**
      2.1. Satisfaction   10
      2.2. Logical implication   12
      2.3. Substructures and reducts   12
      2.4. Composite structures   13

   3. **An example: a mutual exclusion protocol**   **15**
      3.1. Proof of the Mutual Exclusion property   21

2. **System executions**   **24**

   1. **Time and interval orderings**   **24**
      1.1. Finiteness conditions   27

   2. **Definition of system executions**   **30**
      2.1. Global time   30
      2.2. Parallel composition of systems   31

   3. **Higher and lower-level events**   **32**
      3.1. Beginning of events   34

   4. **Specification of registers and communication devices**   **34**
      4.1. Specifying communication devices   37

   5. **Achilles and the Tortoise**   **41**
      5.1   Concurrency   44
      5.2. Zeno's paradox   46

3. **Semantics of concurrent protocols**   **48**

   1. **A protocol language and its flowcharts**   **49**
      1.1. Serial and concurrent procedures   50

v

    1.2.   Flowcharts of protocols   52
    1.3.   States and transitions   55

  2. **Semantics of flowcharts**   57
    2.1   Histories and executions of flowcharts   57
    2.2.   External semantics   59

  3. **Histories as structures**   59

  4. **The Pitcher/Catcher example**   61

4. **Correctness of protocols**   69

  1. **Mutual exclusion revisited**   69

5. **Higher-level events**   78

  1. *Pitcher/Catcher* **revisited**   79

  2. **Procedure calls**   90

  3. **KanGaroo and LoGaroo**   91
    3.1.   Formalizing the proof   95

  4. **The Aimless Protocols**   99
    4.1.   Higher-level relations   101

  5. **Local registers**   106

PART 2
Shared-variable communication

6. **On the Producer/Consumer problem: buffers and semaphores**   111

  1. **The Producer/Consumer problem**   111

  2. **Buffer cells**   114

  3. **Semaphores**   115
    3.1.   The textbook specification   117
    3.2.   Abstract specification of semaphores   122

  4. **Load/unload with semaphores**   126

  5. **A Multiple Process Mutual Exclusion Protocol**   129

7. **Circular buffers**   132

  1. **Unbounded sequence numbers**   132
    1.1.   The function *activate*   136
    1.2.   Safety   140
    1.3.   Liveness   141

  2. **Bounded sequence numbers**   142

PART 3
Message communication

8. Specification of channels 147

   1. Channels 148
     1.1. Capacity of a channel 151

   2. A redressing protocol 152

   3. Sliding Window Axioms 154

   4. A Multiple Producer Protocol 158

9. A sliding window protocol 164

   1. Protocol analysis and definition and higher-level events 170
     1.1. Higher-level functions 177

   2. The Sliding Window Axioms hold 184
     2.1. Complete *SEND/RECEIVE/ACK* events 184

   3. Correctness of the protocol 187

10. Broadcasting and causal ordering 189

   1. *Send/Receive* Network Signature 190

   2. Message Domain · 191

   3. Causality 193

   4. Causality preservation and deliveries 196
     4.1. Uniform deliveries 198

   5. Time-stamp vectors 199

11. Uniform delivery in group communication 203

   1. A generic Uniform Delivery Protocol 203
     1.1. The Time-Stamp Vector Axioms are satisfied 207
     1.2. Data Buffers 209
     1.3. The All-Ack Protocol 212

   2. Correctness of the generic protocol 213

   3. The Early Delivery Protocol 215

   4. A worked-out example 221

Epilogue: Formal and informal correctness proofs   232

References   234

Index   236

Preface

The title 'Models for Concurrency' indicates that models (in the sense of model theory) are applied here to the analysis of concurrent protocols, and not that the diverse paradigms for modeling concurrency are explained.

The dominating tenet in concurrency research today takes the notion of a state as the basic unit, and uses it to describe systems and processes' behavior. A state is an assignment of values to a set of variables, and analysis of a system involves defining this set of variables and determining the rules-of-change that relate states to their successors. A 'history' is a sequence of states, possibly infinite, governed by these rules. Correctness proofs are essentially proofs 'by induction', whereby an assertion is proved to hold in each state in a history if it holds in the initial state and then the rules of change will not falsify it. I find that decomposing reality as a sequence of states is often inadequate and unnatural because we usually think of events − not of states − in forming our view of reality. These events overlap in time in such a complex manner that the state rendering of reality often becomes artificial.

When thinking and arguing informally about concurrent systems, computer scientists often use what has been called behavioral reasoning ("before sending the message, process P read register R and realized that ..."). This type of reasoning was found by many to be unreliable, and formal methods were suggested for proving protocols' correctness. For example, L. Lamport and F. B. Schneider wrote ([21] page 203):

> Most computer scientists find it natural to reason about a concurrent program in terms of its behavior − the sequence of events generated by its execution. Experience has taught us that such reasoning is not reliable; we have seen too many convincing proofs of incorrect algorithms. This has led to assertional proof methods, in which one reasons about the program's state instead of its behavior. Unlike behavioral reasoning, assertional proofs can be formalized − i.e., reduced to a series of precise steps that can, in principle, be machine-verified.

I agree that it is natural to reason about a concurrent program in terms of its behavior, but I believe that behavioral reasoning can be formalized too. Using first-order sentences and models that describe events and their attributes, one can transform behavioral reasoning into a formal mathematical proof and still keep the main features of the informal argument. The purpose of the book is to describe this approach and its background with many examples.

The basic mathematical notion used here is that of a model for a first-order theory. The first chapter introduces the required concepts from model theory (only the very elementary concepts are needed). The second chapter concentrates on system executions: the structures employed here to model concurrency. The third chapter, the heart of the book, describes the semantics of a simple language that allows concurrent executions of sequential programs. Chapter 4 explains how this semantics can be used to prove the correctness of protocols, and Chapter 5 deals with the question of resolving executions into higher-level and lower-level granularities.

The second part of the book explains the producer/consumer problem and describes two (known) solutions. The first, in Chapter 6, uses semaphores, and the second, in Chapter 7, uses a circular buffer.

The third part of the book deals with message passing rather than shared memory. The first chapter there specifies some notions connected with channels (such as non-lossy channels) which are used in the following chapters, and introduces an axiomatic treatment of sliding window protocols. Chapter 9 picks up the producer/consumer problem again, but now with messages through a channel. The purpose of this chapter is to prove the correctness of the Sliding Window Protocol. Chapter 10 discusses basic concepts (such as causality) in networks. Chapter 11 is about uniform deliveries; it is centered around a very interesting protocol: the Early Delivery Protocol of Dolev, Kramer, and Malki [14].

Not everybody is interested in completely formal correctness proofs that are written out according to strict first-order rules. The designers of a system or the programmer seeking to improve her algorithm mostly want to be convinced that the protocol is correct and to find precise descriptions of its behavior. The second and third parts of the book are written in everyday mathematical parlance that can 'in principle' be fully formalized. This approach can be valuable to those who wish to express concurrency arguments in a rigorous language (that of model theory) but do not desire to transform them into a fully formal first-order proof. (It will take more work to provide tools that help to make such transformations – an important issue though beyond the scope of this book and its methods.)

These notes originated with a course given in the summer of 1995 at Ben-Gurion University to a group of students from the industry, whose definite expectations and interests largely determined the form and content of my lectures. More practical examples were required to keep the students' interest, and no obvious background could be assumed. I am grateful to all of them. including the one who kept asking why so much effort is invested in proving obvious facts.

The book was typeset with the AMS LATEX program. I wish to express my admiration to the generosity of those who developed such programs and made them available to all. I was lucky to have two experts among my friends, James Cummings and Martin Goldstern, who knew how to adapt the program to my needs. I am also grateful to the staff of Gordon and Breach for editorial help and cooperation. Additional thanks go to Bob Constable who has made valuable suggestions concerning Chapter 2.

PART 1

Semantics of distributed protocols

The first part introduces some elementary notions of model theory and uses them to explain the semantics of concurrently communicating processes. We formally define what an execution of a protocol is and what it means to prove that a protocol satisfies some correctness statements.

1
Elements of model theory

First-order languages and their interpretations (models) are used in this book to specify systems and describe the behavior of programs. This chapter introduces some of the elementary notions from model theory that are needed, such as structure, first-order language, and the satisfaction relation. We study these notions informally and only to such a degree as required to read the book.

1. Structures

In this chapter we shall be concerned with language and models (structures)— two interrelated concepts that play a central role in logic and in this book. To define a language its signature is needed first.

DEFINITION 1.1 (SIGNATURE). *We say that L is a signature if L is a four-tuple $\langle \overline{P}, \overline{F}, \overline{c}, arity \rangle$ where:*

(1) *$\overline{P}$ is a finite sequence $\langle P_1, \ldots, P_k \rangle$ of symbols called predicates. For each predicate P_i in the sequence, $arity(P_i)$ is a non-zero natural number called "the arity of P_i".*

(2) *$\overline{F} = \langle F_1, \ldots, F_\ell \rangle$ is a finite sequence of "function symbols", and, again, a non-zero natural number $arity(F_i)$ is associated with each function symbol.*

(3) *$\overline{c} = \langle c_1, \ldots \rangle$ is (a possibly infinite) sequence of "constants".*

For example, to describe the natural numbers, a signature L may be formed by taking a single predicate $<^*$ with arity 2, two function symbols, $+^*$ and $\times^*$, with arity 2, and a single constant, the symbol 0^* (it would also be natural to add all constants n^* to represent the natural numbers $n \in \mathbb{N}$). L is thus just a set of symbols with associated arities. The asterisk, for example $+^*$, is to emphasize that this is just a symbol rather than the familiar addition operation $+$. In later chapters I will not be that careful and the reader will have to find the status of the symbol from the context. The standard interpretation for this signature is obtained by taking as the universe of discourse the set of natural numbers $\mathbb{N}$, the familiar ordering relation as an interpretation of $<^*$, the addition and multiplication operations as interpretations of $+^*$ and $\times^*$, and the number zero to interpret 0^*.

Another natural interpretation for L is obtained by taking the real numbers. But arbitrary interpretations are also possible: Indeed any non-empty set with a binary relation, two functions, and a constant can interpret L. Of course the choice of a symbol may indicate the intentions of the users, but these intentions cannot replace a definition.

We are now going to define interpretations in general.

DEFINITION 1.2 (INTERPRETATION). *Let* $L = \langle \overline{P}, \overline{F}, \overline{c}, arity \rangle$ *be a signature.* $\mathcal{M}$ *is called an interpretation (or a structure) for* L *if* $\mathcal{M} = \langle A, \overline{P}^{\mathcal{M}}, \overline{F}^{\mathcal{M}}, \overline{c}^{\mathcal{M}} \rangle$ *consists of the following.*

(1) *A is a non-empty set denoted $|\mathcal{M}|$ and called the* universe *of $\mathcal{M}$. Members of A are called "individuals" of $\mathcal{M}$.*

(2) *$\overline{P}^{\mathcal{M}} = \langle P_1^{\mathcal{M}}, \ldots, P_k^{\mathcal{M}} \rangle$ associates with each predicate P_i in L of arity $m = arity(P_i)$ an m-ary relation $P_i^{\mathcal{M}}$ on $|\mathcal{M}|$. That is $P_i^{\mathcal{M}} \subseteq A^m$. (For any set A, A^m denotes the collection of all m-tuples from A.) In particular, every unary predicate symbol is associated with a subset of the universe.*

(3) *$\overline{F} = \langle F_1^{\mathcal{M}}, \ldots, F_\ell^{\mathcal{M}} \rangle$ is an interpretation of all function symbols. For each function symbol F_j, of arity m, $F_j^{\mathcal{M}}$ is an m-place function*

$$F_j^{\mathcal{M}} : |\mathcal{M}|^m \to |\mathcal{M}|.$$

(The notation $f : X \to Y$ means that f is a function from X to Y. Thus $F_j^{\mathcal{M}}$ is defined on $|\mathcal{M}|^m$, the set of m-tuples of individuals, and takes values in $|\mathcal{M}|$.)

(4) *Finally, $\overline{c}^{\mathcal{M}} = \langle c^{\mathcal{M}} \mid c$ a constant$\rangle$ interprets the constants of L: For every constant c in L, $c^{\mathcal{M}}$ is an individual of $\mathcal{M}$. That is, $c^{\mathcal{M}} \in |\mathcal{M}|$.*

Reality often forces us to consider objects of different natures. There are two possible ways to model two (or more) sorts of individuals in a structure. The first is to assume that there is a single universe containing a mixture of individuals of all sorts, and to distinguish them by different predicates. The second is the way adopted here: to assume not just a single universe of discourse, but several domains, called sorts, in a single structure. Such structures are said to be multi-sorted. I give an example for this in the following subsection, but multi-sorted signatures are defined first.

DEFINITION 1.3 (MULTI-SORTED SIGNATURE). *A multi-sorted signature is a sequence of the form*

$$L = \langle S_1, \ldots, S_n; \overline{P}, \overline{F}, \overline{c}, arity, sort \rangle$$

where $\langle \overline{P}, \overline{F}, \overline{c}, arity \langle$ is a signature, $S_1, \ldots, S_n$ is a list of symbols called sorts, and sort is a function that associate with each predicate, function, or constant its sort as follows.

If P is a predicate of arity k, then $sort(P)$ is a k-tuple of sorts. That is $sort(P) = \langle X_1, \ldots, X_k \rangle$ where each X_i is some S_j (repetitions are allowed).

The intention is that the question of whether $P(x_1, \ldots, x_k)$ holds or not can be asked only if each x_i is of sort X_i. Similarly, with each k-ary function symbol F, sort(F) is a k+1 tuple of sorts defining the sorts both of the domain and the range of F. Finally sort(c) for a constant c gives the sort of c. We often introduce a semicolon (;) after the list of sorts to separate them from the predicates and functions.

DEFINITION 1.4 (MULTI-SORTED INTERPRETATION). *An interpretation $\mathcal{M}$ for a multi-sorted signature L consists of*

(1) *a universe $|\mathcal{M}| = S_1^{\mathcal{M}} \cup S_2^{\mathcal{M}} \cdots \cup S_n^{\mathcal{M}}$, which is now a union of the sorts (not necessarily a disjoint union), and*
(2) *the predicates, function symbols and constants, which are interpreted in accordance with their sorts.*

This means, for example, that if F is a two-place relation symbol and sort(F) = $\langle S_1, S_2 \rangle$, then $F^{\mathcal{M}} \subseteq S_1^{\mathcal{M}} \times S_2^{\mathcal{M}}$. Similarly, if g is a unary function with sort(g) = $\langle S_1, S_2 \rangle$, then $g^{\mathcal{M}} : S_1^{\mathcal{M}} \to S_2^{\mathcal{M}}$.

DEFINITION 1.5 (ISOMORPHISM). *Let L be a signature with sorts $S_1, \ldots, S_n$, and let $\mathcal{M}_1, \mathcal{M}_2$ be two interpretations of L. We say that $f : |\mathcal{M}_1| \to |\mathcal{M}_2|$ is an isomorphism of $\mathcal{M}_1$ and $\mathcal{M}_2$ if f is one-to-one from the universe of $\mathcal{M}_1$ and onto the universe of $\mathcal{M}_2$ such that:*

(1) *For every sort S_i, $m \in S_i^{\mathcal{M}_1}$ iff $f(m) \in S_i^{\mathcal{M}_2}$. (iff stands for "if and only if".)*
(2) *For every n-ary predicate symbol P and n-tuple $m_1, \ldots, m_n \in |\mathcal{M}_1|$,*

$$\langle m_1, \ldots, m_n \rangle \in P^{\mathcal{M}_1} \text{ iff } \langle f(m_1), \ldots, f(m_n) \rangle \in P^{\mathcal{M}_1}.$$

(3) *For every n-place function symbol F,*

$$f(F^{\mathcal{M}_1}(m_1, \ldots, m_n)) = F^{\mathcal{M}_2}(f(m_1), \ldots f(m_n)).$$

Similarly, if c is any constant then $f(c^{\mathcal{M}_1}) = c^{\mathcal{M}_2}$.

Since we shall identify two isomorphic structures, this concept reflects our understanding that the inner composition of the elements of the universe of a structure is not relevant. These elements are just abstract "points" devoid of any inherent meanings, were it not for the functions and predicates defined on them.

1.1. Examples of multi-sorted structures. In the first example we model the situation in which several measurements of light intensity are made by some light-meter. A measurement evaluates the average intensity of light during the opening of the meter, which is by definition the integral of the intensity over the aperture interval divided by its length. We want to be able to express in our language the fact that distinct measurements may be made over different exposure intervals and may show different values. For this, every structure contains a set of "events" that represent measurements, and a value is assigned to each event to represent the result of the measurement. Hence two sorts of individuals are needed: measurement events and real numbers (if we want the results to be real numbers). Moreover, to evaluate the measurement error, we would like to have the "true" light intensity function in the structure, that is the function that gives the real intensity at each instant. We shall denote this function by I and let $I(t)$ denotes the light intensity at moment t. We also want the integral function of I. This integral function will allow us to express the average intensity.

In our example the measurement events are one sort and the real numbers are another. An advantage of this two-sorted approach is that functions and relations may be specific to certain sorts. The addition operation, for example, is defined on the real numbers, and it is meaningless to ask for the addition of events.

So we define first a signature K that contains two sorts: E and R (the elements of E are called events and those of R numbers). In addition, K contains the following.

(1) A binary predicate $<$ and binary function symbols $+$, $-$, $\times$, $/$ over sort R. (For typographical clarity, we no longer use the asterisks.) When we say that $+$, for example, is over R we mean that $sort(+) = \langle R, R, R \rangle$; that is to say that $+$ is interpreted as a function taking pairs of individuals of sort R into individuals of sort R.

(2) For each rational number q, a constant q.

(3) Two function symbols, $Left\_End$ and $Right\_End$, are defined on E and give values in R. (The intention is to use them to give for every event e its temporal interval $(Left\_End(e), Right\_End(e))$).

(4) A unary function $Value$ defined on E and giving values in R. (The intention is to use it for the values of the measurement events.)

(5) Two unary function symbols I and A. $I(t)$ gives the true light intensity at time t, and $A(t)$ is the integral of I from some start-up time t_0 to t (so t_0 is a constant in this signature). A more familiar notation for $A(t)$ would be $\int_{t_0}^{t} I(x)dx$, but of course it is too far from our syntax.

A standard real number interpretation $\mathcal{M}$ of K is a structure containing the real numbers $\Re$ (interpreting the sort R) and a set $E^{\mathcal{M}}$ of "events". The interpretation of the arithmetical operations is the standard interpretation on $\Re$ (for example, $<^{\mathcal{M}}$ is the natural ordering on $\Re$). For each individual e in $E^{\mathcal{M}}$, $Left\_End^{\mathcal{M}}(e)$, $Right\_End^{\mathcal{M}}(e)$, $Value^{\mathcal{M}}(e)$ are real numbers with $Left\_End^{\mathcal{M}}(e) <^{\mathcal{M}} Right\_End^{\mathcal{M}}(e)$. The interpretation of I is an arbitrary (integrable) real-valued function $I^{\mathcal{M}}$, and $A^{\mathcal{M}}$ is its integral (from time $t_0^{\mathcal{M}} \in \Re$).

Of course there is nothing in the symbols to force this particular interpretation, and many others are possible. This example is brought here just to give some idea of the diversity of situations where multi-sorted structures can be used, and though it is used again in the next section it will not be used later on.

Two-sorted structures are useful to handle pairs and finite sequences, as the second example shows. This will be significant for later development in the book. Usually we write pairs with angled brackets $\langle a, b \rangle$, but for simplicity of expression we often use round brackets (a, b).

In set theory one learns how to form pairs as sets. A pair $\langle a, b \rangle$ can be defined as a set $\{\{a\}, \{a, b\}\}$. We do not use this (or any other) particular representation here as only the abstract properties of pairs and finite sequences are needed. We shall define now the language in which these abstract properties are expressed.

We first define the pair signature. There are two sorts, U (for the set of individuals from which pairs are formed) and P (for the pairs). There is a binary function symbol *make_pair* defined on U and returning values in P, and there are two unary function symbols *first* and *second* defined on P and returning values in U.

Two axioms determine the properties of these pairing and unpairing functions. Let us agree that variables x and y vary through U and variables p and p' vary through P.

(1) $\forall x, y \, [first(make\_pair(x,y)) = x \wedge second(make\_pair(x,y)) = y]$

(2) $\forall p, p' \, [first(p) = first(p') \wedge second(p) = second(p') \longrightarrow p = p']$

Observe that we did not have to say in these axioms that $make\_pair(x,y)$ is in P, as this is a consequence of the signature requirement which says that the sort of $make\_pair$ is $\langle U, U, P \rangle$, namely that $make\_pair$ is a binary function from U to P. An advantage of using a multi-sorted language is that each variable is connected with a certain fixed sort, and this brings some economy in writing formulas. For example, a longer form of the first axiom would be $\forall x, y \in U \ldots$, but here there was no need to qualify the range of x and y which were assumed to be U-variables. If the traditional notation $\langle x, y \rangle$ for the expression "$make\_pair(x,y)$" is used, then the first axiom becomes $\forall x, y(first(\langle x, y \rangle) = x \ldots$.

We did not require that sorts U and P are disjoint. In many cases it is natural to add $\forall x \in U(x \notin P)$, but there are occasions when we want $P \subset U$, which enables forming pairs $\langle x, y \rangle$ when x or y (or both) are themselves pairs.

If $\mathcal{M}$ is an interpretation of the signature defined above that satisfy the two axioms, then $U^{\mathcal{M}}$ and $P^{\mathcal{M}}$ are just sets, and the fact that we shall call members of $U^{\mathcal{M}}$ "elements" and members of $P^{\mathcal{M}}$ "pairs" does not say anything about their structure. In fact, the individuals of $\mathcal{M}$ are devoid of any structure and only the functions and predicates defined over them can convey a meaning.

Let $\mathcal{M}$ be an interpretation for this pairing/unpairing signature, and assume that $P^{\mathcal{M}} \subset U^{\mathcal{M}}$. Then a notation $\langle a_0, \ldots, a_{n-1} \rangle$ can be introduced for $n \geq 2$ and for a_i's in $U^{\mathcal{M}}$ as follows: If $n > 2$ then

$$\langle a_0, \ldots, a_{n-1} \rangle = make\_pair(\langle a_0, \ldots, a_{n-2} \rangle, a_{n-1}).$$

There are other possibilities to model tuples, and I do not want to suggest that this is the best way: the reader can keep his favorite definition if he wants to, and I will avoid referring to a particular representation by adhering to the following convention. Whenever a structure $\mathcal{N}$ is mentioned and $U^{\mathcal{N}}$ is a specific sort in $\mathcal{N}$, if I say that $\mathcal{N}$ is *equiped with pairs*, I mean that there is in fact another sort in $\mathcal{N}$, a sort of pairs of members of $U^{\mathcal{N}}$, even though this sort and the related functions were not specified beforehand. Since pairs are ubiquitous, it is reasonable to assume that all our structures are equiped with pairs.

Our last example concerns representation of functions as individuals in a structure (rather than as interpretations of function symbols). Suppose V and A are sets (the objects of V will be called variables and those of A values) and let S be the collection of all functions $s : V \to A$ (members of S will be called states). I

want to form a structure in which the functions of S can be discussed, and I am
going to describe two possible ways of doing it: both are natural and the reader
may prefer the first alternative in certain situations and the second in others.

One possibility is to gather all three sets V, A, and S as individuals in a
three-sorted structure $\mathcal{S}$. If s is of sort S, v of sort V, and a of sort A, then the
expression $s(v) = a$ cannot be formed since s (being an individual in a structure)
is not a function: it is devoid of any internal structure. (A first-order expression
$F(v)$ can only be formed when F is a function symbol in the relevant signature.)
One therefore needs a binary function symbol $apply$ in the signature of $\mathcal{S}$ and
then $apply(s, v)$ takes the place of $s(v)$. Having said this, one can relax the
graphical representation of formulas, and declare that the parentheses (,) form a
special symbol and so $s(v)$ is, by definition, a nicer representation of $apply(s, v)$.

Another possibility is to have only S and A as sorts of $\mathcal{S}$, and to introduce to
the signature a special function symbol Val_v for every variable $v \in V$. It would
be most natural to do so if V is a small finite set. Then $Val_v(s)$ represents the
value of s at argument v.

2. First-order languages and satisfaction

In this section we define the notions of (first-order) language and satisfaction
to a degree required for reading this book, but without giving a full exposition.

To prepare the ground for a mathematical definition we return to the light
intensity measurement from Section 1.1. Suppose that we want to say about a
light-meter that it guarantees that every measurement is accurate to within 0.1%
of the real average intensity, provided its exposure interval is at least 20 seconds
but not more than 35. I will write this statement three times, in increased
degrees of formality.

(1) For every measurement event e, if the duration of e is between 20 seconds
    and 35 seconds, then the value returned by e is between $0.999 \times r$ and
    $1.001 \times r$, where r is the real average intensity which is obtained by dividing
    the intensity integral over the measurement period by the duration of this
    period.

(2) For $r = (A(t_2) - A(t_1))/(t_2 - t_1)$, $t_1 = Left\_End(e)$, and $t_2 = Right\_End(e)$
    the following holds:

$$\forall e \in E \; [(20 \leq duration(e) \leq 35) \to 0.999 \times r < Value(e) < 1.001 \times r].$$

(3) There are two problems in trying to write this in "pure" first-order lan-
    guage: First, the function $duration$ and the relation $\leq$ are not in our
    language, and second there are no facilities for using expressions such
    as "where r is etc." The first problem can either be solved by adding
    those symbols to our language, or by showing that they are express-
    ible (definable) in it. The expression $a \leq b$ for example can be re-
    placed by $(a < b \lor a = b)$, and $20 \leq duration(e)$ can be replaced by
    by $20 \leq Right\_End(e) - Left\_End(e)$.
    The second problem is not solved elegantly: One can either substitute
    $(A(t_2) - A(t_1))/t_2 - t_1$ in every occurrence of r (which results in complex

expressions), or use existential quantifiers thus:

$$\forall e \in E \ (\exists t_1, t_2, r \in R)$$
$$(t_1 = Left\_End(e) \wedge t_2 = Right\_End(e) \wedge r = (A(t_2) - A(t_1))/t_2 - t_1) \wedge$$
$$(20 \leq t_2 - t_1 \leq 35 \longrightarrow 0.999 \times r < Value(e) < 1.001 \times r).$$

The reader can appreciate now why I seldom write the formal sentences in full, but prefer an informal discourse.

The expression $\forall e \in E \ \phi$ used above as a formalization of "for all events e ϕ" can be written simply as $\forall e \ \phi$ if we agree that variable e varies only over events. In general we adopt this convention of using special variables for each sort of objects, but the qualified quantification $\forall x \in E \ \phi$ is still used here and there. (Many authors use expressions such as $\forall e : E$, which is very reasonable because the relationship between a variable and its sort is not exactly the membership relation.)

We leave our examples now and define first-order languages in general. In addition to the signature symbols, logical symbols and variables are needed to construct sentences.

DEFINITION 2.1. *A (first-order) language consists of the following items.*

(1) *A multi-sorted signature* $\langle S_1, \ldots, S_n, \overline{P}, \overline{F}, \overline{c}, arity, sort \rangle$.
(2) *A collection* V_L *of variables. Each variable* $X \in V_L$ *is associated with a specific sort* $sort(X) = S_i$. *(So sort is not only defined on the* $\overline{P}, \overline{F}, \overline{c}$ *symbols, but on the variables as well.)*
(3) *A symbol is needed for each logical connective. We use* $\wedge$ *for conjunction,* $\vee$ *for disjunction,* $\neg$ *for negation, and* $\longrightarrow$ *for implication.*
(4) *Quantifiers:* $\forall$ *is the universal quantifier "for all", and* $\exists$ *is the existential quantifier "there exists".*
(5) *Parentheses are also needed. We use* $)$ *and* $($.

Thus the difference between a signature and a language is that the latter also contains the logical symbols and a list of variables in each sort.

Now there are *formation rules* by which terms and formulas can be constructed. First terms are defined. In the context of a multi-sorted logic each term is of a particular sort:

(1) Any constant or variable t is a term, and its sort is given by $sort(t)$.
(2) If F is a function symbol of arity n and $\tau_1, \ldots, \tau_n$ are terms then $F(\tau_1, \ldots, \tau_n)$ is a term of sort T if the following holds:

$$sort(F) = \langle S_1, \ldots, S_n, T \rangle, \quad \text{and each } \tau_i \text{ is of sort } S_i.$$

That is, if F is specified to take arguments from sort S_i at its ith coordinate and to return arguments in sort T, then each τ_i must be of sort S_i, and the term $F(\tau_1, \ldots, \tau_n)$ is of sort T.

Next formulas are defined; first atomic and then compound formulas. An atomic formula is any sequence of symbols of the form $P(\tau_1, \ldots, \tau_n)$ where P is a predicate of arity n and each τ_i is a term of the appropriate sort (which sorts can be in the ith coordinate is determined by *sort*(P)). The equality predicate $=$ exists in every signature (but the tradition is to write $\tau_1 = \tau_2$ rather than $= (\tau_1, \tau_2)$).

The logical connectives and quantifiers are used to construct compound formulas as follows:

(1) Any atomic formula is a formula.
(2) If ϕ_1 and ϕ_2 are formulas, then so are
   - $(\phi_1 \wedge \phi_2)$
   - $(\phi_1 \vee \phi_2)$
   - $(\phi_1 \longrightarrow \phi_2)$
   - $(\neg\phi_1)$
(3) If ϕ is a formula and x a variable then $(\forall x \phi)$ and $(\exists x \phi)$ are formula. If x is a variable of sort S, then $\forall x \in S\phi$ and $\exists x \in S\phi$ are also formulas, and the redundant symbolism is only used for the reader's convenience.

In the definition of terms given above we assume that the function symbols are written in the prefix notation whereby a function always precedes its terms. Yet in our previous light-meter example $(A(t_2) - A(t_1))/(t_2 - t_1)$ is a term composed of both prefixed and infixed symbols. There are standard natural ways of dealing with this issue which we shall not discuss.

As we defined it, a language is a list containing its signature, variables, etc. However, the term "language" often refers to the set of all terms and formulas as defined above. Thus, we shall say "let ϕ be a formula in L" etc. instead of the pedantic "let ϕ be formed from the symbols of L according to the formation rules." The distinction between bound and free occurrences of variables in a formula is very important. An example can clarify this distinction. In the formula

$$(\forall x R(x, y)) \vee A(x)$$

y is a free variable, x has three occurrences: it is bound in the second, and free in the third (the one with $A(x)$). The first occurrence is a formal modifier of the universal quantifier. A formula says something about its free variables, but nothing at all about its bound variables. For example $\exists y(2y = x)$ says that x is an even number, but doesn't say anything about y. (A formal definition of free and bound occurrences can be found in any logic textbook.) If ϕ is a formula with free variables x_1, y_3 and z for example, then we shall express this by writing $\phi(x_1, y_3, z)$. A *sentence* is a formula with no free variables.

The term "first-order" refers to the fact that quantification is over individuals rather than classes (or functions). We shall not consider here higher-level languages, and we say "first-order language" as opposed to "computer-language".

2.1. Satisfaction.

DEFINITION 2.2 (ASSIGNMENT). *Let L be a first order language, V its set of variables, and S a structure for L. An assignment is a function σ from V (or a subset of V) into the universe of S. It must respect the sorts of the variables. That is, if x is a variable of sort V in L, then $\sigma(x)$ is in V^S.*

Given an assignment σ, any term of the language constructed with variables that are in the domain of σ acquires a value in its sort. For example, the value of $F(x_2, z, c)$ is $F^S(\sigma(x_2), \sigma(z), c^S)$. Here F^S is the interpretation of F as a three-place function on the universe of S, c^S is the interpretation of the constant c as an individual in S, and $\sigma(x_2)$, $\sigma(z)$ are the values of these variables under σ. In general, if τ is any term then τ^σ denotes the value of τ under σ (the interpretation S being fixed).

If ϕ is a formula and σ an assignment to S, then the relation

$$S \models \phi[\sigma]$$

means that ϕ is true in S when we substitute for each free variable x of ϕ the value $\sigma(x)$. Relation $\models$ is called the satisfaction relation, and we say that ϕ is satisfied in S under assignment σ. We can restrict our discussion to finite assignments if, whenever $S \models \phi[\sigma]$ is written, it is clear that σ is defined on all the free variables of ϕ.

The definition of $\models$ is due to A. Tarski and was a major achievement in logic; the reader is encouraged to see how it is formally done in any logic textbook. We will only state the inductive requirements that this relation must satisfy.

(1) For any atomic formula of the form $R(\tau_1, \ldots, \tau_n)$

$$S \models R(\tau_1, \ldots, \tau_n)[\sigma] \quad \text{iff} \quad \langle \tau_1^\sigma, \ldots, \tau_n^\sigma \rangle \in R^S.$$

For example, if S is the standard structure of the integers and R is a two-place predicate interpreted as the $<$ relation, then $R(x, y)$ holds under assignment σ iff $\langle \sigma(x), \sigma(y) \rangle \in R^S$, or simply iff $\sigma(x) < \sigma(y)$.

(2) $S \models \neg\phi[\sigma]$ iff it is not the case that $S \models \phi[\sigma]$.

(3) $S \models (\phi \wedge \psi)[\sigma]$ iff both $S \models \phi[\sigma]$ and $S \models \psi[\sigma]$. Similar statements can be formed for all other logical connectives.

(4) $S \models \exists x \phi[\sigma]$ iff there exists some assignment σ' such that
   - for every variable y different from x $\sigma'(y) = \sigma(y)$,
   - $\sigma'(x)$ is defined, and
   - $S \models \phi[\sigma']$.

Informally we can simply say that $\exists x \phi$ is satisfied iff there exists an individual a of the appropriate sort in the universe of S so that when a is substituted for x in ϕ then the resulting formula is satisfied in S. Note however that formulas are syntactical entities, assignments are functions, and a substitution of a value in a formula has not been defined.

(5) In a similar way, $S \models \forall x \phi[\sigma]$ iff for any assignment σ', if $\sigma'(x)$ is defined and σ' is equal to σ on each variable different from x, then $S \models \phi[\sigma']$.

The relation $S \models \phi[\sigma]$ depends only on the values $\sigma(x)$ of the variables that are free in ϕ. In particular, if ϕ is a sentence (a formula with no free variables) then the assignment can be omitted and we may write $S \models \phi$.

Sometimes we write the assignment function σ explicitly: for example if $\sigma(x_i) = a_i$ then we can write

$$S \models \phi[x_1, \ldots, x_n \; / \; a_1, \ldots, a_n].$$

We may even substitute individuals for variables and write an impure formula such as $3 < 4$ instead of $x < y[x, y \; / \; 3, 4]$, when clarity is preferred to rigor. (Of course, if 3 and 4 are constants in L then there is nothing wrong with $3 < 4$, only when they are individuals (natural numbers) is there a danger of confusing two types of entities.)

DEFINITION 2.3 (MODEL). *We say that a* collection *of sentences T in some language is satisfied by a structure $\mathcal{M}$ if every sentence in T is satisfied by $\mathcal{M}$; we then say that $\mathcal{M}$ is a model of T. We write this as $\mathcal{M} \models T$. A collection of sentences that has a model is called a theory* .

2.2. Logical implication. When proving properties of a program, we will follow three steps: (1) express these properties as first-order sentences, (2) define a family S of structures that describe the program's executions, (3) prove that the sentences hold in every structure from S. These correctness proofs have the form $\Phi \implies \psi$ where Φ is a set of sentences defining the structures and ψ is a single sentence expressing the correctness conditions. We therefore have to define exactly what $\Phi \implies \psi$ means, and how one proves such implications.

Fix a language (and signature) L. Suppose that Φ and Ψ are two sets of sentences in L. Then $\Phi \implies \Psi$ means that for every structure $\mathcal{M}$ of L, if $\mathcal{M}$ is a model for Φ then it is also a model for Ψ. In case $\Psi = \{\psi\}$ is a singleton we write $\Phi \implies \psi$ rather than $\Phi \implies \{\psi\}$. The relation $\Phi \implies \Psi$ is also meaningful when Φ and Ψ are sets of formulas (which may contain free variables). This relation means then that for every structure $\mathcal{M}$ of L and every assignment σ in $\mathcal{M}$, if $\mathcal{M}$ (under σ) satisfies every formula in Φ then it also satisfies every formula in Ψ. Again, for singletons, we write $\phi \implies \psi$ for simplicity.

When proving $\Phi \implies \psi$, we really show that ψ is *derivable* from Φ. The notion of derivation can be defined mathematically, and this leads to the notion of a first-order proof. We will not do this however and describe our proofs in the usual everyday language of mathematics.

In the following section we shall give an example where $\Phi \implies \psi$ is proven relative to a fixed family S of interpretations. That is, S is a collection of interpretations of L, and $\Phi \implies \psi$ means in this context that for every structure $\mathcal{M} \in S$ that is a model of Φ, $\mathcal{M} \models \psi$. When proving that $\Phi \implies \psi$ holds for S, we use the special properties of the models in S. These are usually not first-order properties and hence the proofs cannot be turned directly into first-order proofs. The experienced reader, however, can formulate the first-order axiom schemes that replace the second-order properties and turn the proof into a first order proof.

2.3. Substructures and reducts. Let L be a first-order language. Given two structures $\mathcal{S}_1$ and $\mathcal{S}_2$ for L, we say that $\mathcal{S}_1$ is a substructure of $\mathcal{S}_2$ iff:

(1) $|\mathcal{S}_1| \subseteq |\mathcal{S}_2|$ (the universe of $\mathcal{S}_1$ is a subset of the universe of $\mathcal{S}_2$, where the universe of a multi-sorted structure is the union of its sorts).

(2) $\mathcal{S}_1$ is obtained by restricting the interpretation of $\mathcal{S}_2$ to the universe of $\mathcal{S}_1$. This means the following:

(a) For every sort T and individual a from $|\mathcal{S}_1|$, $a \in T^{\mathcal{S}_1}$ iff $a \in T^{\mathcal{S}_2}$.

(b) For every predicate P of arity n in L and for every n-tuple of individuals from $|\mathcal{S}_1|$,

$$\langle a_1, \dots, a_n \rangle \in P^{\mathcal{S}_1} \text{ iff } \langle a_1, \dots, a_n \rangle \in P^{\mathcal{S}_2}.$$

(c) For every function symbol F, F^{S_1} is the restriction of F^{S_2} to $|S_1|$.
(The restriction of a function $f : A \to B$ to a subset of its domain
$A' \subseteq A$ is the function $f' : A \to B$ defined by $f'(x) = f(x)$ for $x \in A'$.
We write $f' = f|A$ for the restriction.)

(d) If c is a constant in L then $c^{S_1} = c^{S_2}$.

We shall see that the notion of a *reduct* is very useful. A signature L_1 is said to be
a sub-signature of L_2 if L_1 is a signature that consists only of sorts, predicates,
function symbols and constants from L_2, but it does not change their arity and
sort. (Since L_1 is a signature, if F is a function symbol in L_1, for example, then
both the domain and range of F are sorts in L_1).

DEFINITION 2.4 (REDUCT). *If L_1 is a sub-signature of L_2 and S_2 is a structure for L_2, then the reduct of S_2 to L_1 is the following structure S_1 obtained by
"forgetting" the symbols not in L_1:*

- *The universe of S_1 is obtained by taking the sorts of S_2 that are in L_1.
(The universe of every structure is the union of its sorts, and so $|S_1|$ is
the subset of $|S_2|$ obtained as $\bigcup \{S^{S_2} : S \text{ is a sort in } L_1\}$).*
- *The interpretation of each predicate, function symbol, and constant of L_1
is the same in S_1 and S_2.*

It is convenient to use the notation

$$S_1 = S_2 | L_1$$

for reducts (little danger of confusion with function restriction). We also say in
this case that S_2 is an expansion of S_1 (to L_2).

2.4. Composite structures. The structures we encounter in our studies
and research are usually not interpretations in the strict model-theoretic definition. Indeed (except for logicians) most of the workers in mathematics and
computer science deal with composite structures involving several different structures whose elements have a certain inner composition. For example, a vector
space is a combination of two structures: a vector group and a field (of scalars),
related through a product operation with special properties. In some sense the
vectors are more important than the scalars which are often left in the background. Even though it is a simple routine to transform vector spaces into flat
first-order structures (as we shall see) there is usually no need for that. The advantage of composite structures such as vector spaces over their model theoretic
counterparts is their hierarchical composition from familiar objects. Thus it is
unthinkable to teach linear algebra through model theory (though, perhaps, it is
not a bad idea to deepen the students' understanding of structures in general).
Yet model theory, by treating all structures uniformly, provides a logical basis for
the notion of truth and of proof. Thus, from our point of view, transformations
of composite structures into their model theoretic presentations (which we call
flattening) is very useful.

The notion of composite structure is used here informally to denote those
mathematical objects whose elements have a prior definition and structure—as
opposed to first-order structures whose points are "flat", that is devoid of further
composition. We shall define (in Chapter 3) flowcharts, histories, and executions

of histories in order to define the semantics of concurrent protocols, and these will all be composite structures. We will have to transform them into first-order structures in order to provide a logical framework within which correctness proofs can be conducted, and therefore experience with such transformation can be helpful. We give here two simple examples of transformations, that of vector spaces and that of finite automata.

A standard definition of vector spaces runs as follows: A nonempty set V is said to be a vector space over a field F if V is an abelian group under operation $+_V$, and there is a product operation $\cdot$ such that for every $\alpha \in F$ and $v \in V$, $\alpha \cdot v \in V$, and such that the following properties hold (and here comes the well-known list which we shall not repeat). Of course, the concepts of field and abelian group are well known at this stage so that the student only needs a few examples to grasp this concept. There are two ways to turn vector spaces into first-order structures (to flatten them as we shall say). The first is to add for each scalar $\alpha \in F$ a function symbol $f_\alpha : V \to V$ with the intended interpretation of $f_\alpha(v) = \alpha \cdot v$. The second approach, for which we give more details, is to work with a two-sorted structure containing field elements and vectors.

The signature therefore contains two sorts: F (for the field elements) and V (for the vectors), binary function symbols $+_F$, $\cdot_F$, $+_V$, $\cdot_{F \times V}$ and constants 0_F, 1_F, 0_V, with the well-known restrictions: for example that $\cdot_{F \times V} : F \times V \to V$, or that $0_V \in V$. Then there is the familiar list of first-order axioms in this signature that characterize vector spaces.

The second example involves automata and the words they generate. Let us repeat some of the basic definitions in the form in which they are usually given. An alphabet is a finite set A of symbols. A string (or word) over A is a finite or infinite sequence of symbols from A. The empty string is denoted by ϵ.

A finite automaton over A is a five-tuple of the form $\mathcal{A} = (A, S, s_0, \delta, F)$ where A is an alphabet, S is a finite set (the set of states of $\mathcal{A}$), $s_0 \in S$ is the *initial state*, $\delta : S \times A \to S$ is the *transition function*, and $F \subseteq S$ is the set of *final states*.

This definition of an automaton is one of the (rather rare) examples where the natural definition of a mathematical object actually is a first-order structure. A and S are its sorts (and then the words "symbols", for the elements of A, and "states", for those of S, should be seen as informal additions to the mathematical definition whose aim is just to convey the intention of the user). s_0 is a constant, δ a function symbol, and F a predicate. Of course we do not claim that first-order sentences could characterize the finite automata, since no such sentences can determine the finiteness of A and S, but this is fine with us.

An "infinite word" is a function $\sigma : \mathbb{N} \to A$. The pair $(\mathcal{A}, \sigma)$ is a composite structure, representing a finite automaton and an infinite word that it is ready to read. While it is a well-defined mathematical entity, $(\mathcal{A}, \sigma)$ is not a structure in the model theoretic sense used here. For example, we shall define below the sequence of states that the automaton takes while it reads σ. This perfectly legitimate definition will not give us a first-order formula $\psi(i, s)$ saying that at stage i of the reading the automaton is in state $s \in S$. The problem, of course, is that the composite structure $(\mathcal{A}, \sigma)$ is not ready to carry first-order definitions. We continue now with defining what it means for an automaton *to read* an infinite word σ. An "instantaneous description" of $\mathcal{A}$ is defined to be a pair $\langle s, a \rangle$ where $s \in S$ and $a \in A$. It is a description of a position of the automaton at state s and with a as the next letter to be read. Any infinite word σ over A defines an infinite sequence $\mu = i(\sigma)$ of instantaneous descriptions (or just positions) defined as follows:

(1) $\mu(0) = \langle s_0, \sigma(0) \rangle$. That is, the first position is with the initial state and the first symbol of σ.

(2) If $\mu(n) = \langle s, a \rangle$ then $\mu(n+1) = \langle \delta(s, a), \sigma(n+1) \rangle$. That is, the transition function determines the next position.

If $\mu(n) = \langle s_n, a_n \rangle$ then $\langle s_n \mid n \in \mathbb{N} \rangle$ is the sequence of states that the automaton goes through as it reads σ. Our aim is to transform this composite definition of $(\mathcal{A}, \sigma, \mu)$ into a first-order structure within which we can define a formula saying, for example, that the automaton went infinitely many times through some state s as it read σ. For this we must flatten $(\mathcal{A}, \sigma, \mu)$ into a structure.

Now μ is an infinite sequence, and so we first ask: what is a sequence? As a set-theoretical object, a sequence $\alpha = \langle a_0, a_1 \ldots \rangle$ is a function on $\mathbb{N}$, or an initial segment of $\mathbb{N}$ (the set of natural numbers), and a_i is then just another notation for $\alpha(i)$. In model theoretic terms, a sequence is a two-sorted structure of the form $(H, A, <, \sigma)$ where H is the sort of indices, interpreted either as the standard set $\mathbb{N}$ of natural numbers or as a finite initial segment of $\mathbb{N}$, A is a set of possible values, $<$ is the natural ordering of H and σ is a function from H into A.

Let us call the required flat structure a *computation*. It would thus have the form

$$\mathcal{C} = (A, S, \mathbb{N}; \delta, +, \sigma, \mu; F, <; s_0, 0, 1)$$

where A is the alphabet, S the sort of states, $\mathbb{N}$ the sort of number indices; $+$ is the addition function on $\mathbb{N}$, $<$ is the ordering on $\mathbb{N}$, and $0,1$ the constants; δ is the transition function, σ the "input" infinite string, and μ the resulting sequence of instantaneous descriptions. For every $i \in \mathbb{N}$, $\mu(i) \in S \times A$, but we do not have to introduce $S \times A$ specifically as a sort of $\mathcal{C}$ since we always assume that our structures are closed under pairs, which means that this sort is automatically there. (See the assumption on page 7.)

Now, suppose we want to say about a computation that the automaton went through state $s \in S$ infinitely many times. Assuming that variables i and i' vary over $\mathbb{N}$ the formula for that would be $\forall i \exists i'(i < i' \wedge \mu(i') = \langle s, \sigma(i') \rangle)$. To give another example, suppose we want to say that aba is not a substring of σ (where a, b are variables over A). The formula for that is $\forall i \neg (a = \sigma(i) \wedge b = \sigma(i+1) \wedge a = \sigma(i+2))$ (where $2 = 1 + 1$ of course).

3. An example: a mutual exclusion protocol

This section analyses Peterson's well known critical section protocol (from [25]) and proves that the mutual exclusion property holds. My main aim in presenting the protocol at this stage is to give an example of the assertional method of proof which is the standard approach to concurrency using global states and transitions. The approach advocated in this monograph is different,

procedure P_1	procedure P_2
repeat_forever	**repeat_forever**
begin	**begin**
(1) non-critical section	(1) non-critical section
(2) $Q_1 := true$;	(2) $Q_2 := true$;
(3) $Turn := 1$;	(3) $Turn := 2$;
(4) **wait_until**	(4) **wait_until**
$\neg Q_2 \vee Turn = 2$;	$\neg Q_1 \vee Turn = 1$;
(5) Critical Section of P_1;	(5) Critical Section of P_2;
(6) $Q_1 := false$	(6) $Q_2 := false$
end.	**end.**

FIGURE 1.1. Peterson's mutual exclusion protocol. Q_1 and Q_2 are initially *false*.

and when it is used to reprove the same mutual exclusion property, the reader will have a fuller perspective on the two approaches.

I will be quite informal in some aspects (such as the semantics of the protocol) that do not relate directly to this introduction—these will be defined in the third chapter.

The critical section problem is to devise a protocol that prevents simultaneous access by two or more processes into some sensitive parts of their programs. For example, a process that sends a file to a printer needs to know that no other process can access the printer during that time. Protocols that guarantee this separation are said to have the mutual exclusion property. An additional property is clearly required: that any process requesting its critical section will be able to access it sooner or later. This desirable "liveness" property is called the lockout freedom property.

The protocol and its flowchart are shown in figures 1.1 and 1.2. The two processes access two boolean registers (shared variables) Q_1 and Q_2, and a $\{1, 2\}$ register *Turn*. The initial value of Q_1 and Q_2 is *false*, and the initial value of *Turn* is unimportant. The **wait_until** instructions at line 3 are interpreted here as "busy wait". This means that the process repeatedly reads into two local variables the Q and *Turn* registers until the required condition holds. The flowchart clarifies this.

Two local variables are used by P_1 in the flowchart: r_1 is boolean and s_1 can hold a value from $\{1, 2\}$. The nodes of the flowchart are labeled with instructions that the process is about to execute. A directed edge (which we call "arrow") represents the event of executing the instruction labeled at the node that the arrow is leaving. For example arrow $(1_1, 2_1)$ represents writing *true* on Q_1. Nodes 5_1 (in the P_1 flowchart) and 5_2 (in the P_2 flowchart) are special because they are labeled with a condition (a formula). If the formula is true at a certain state then the arrow labeled with *true* is followed (leading to the critical section) but otherwise the arrow labeled with *false* is taken, which indicates that another round of reading registers is required.

An execution of this protocol is described as a sequence of states whose first

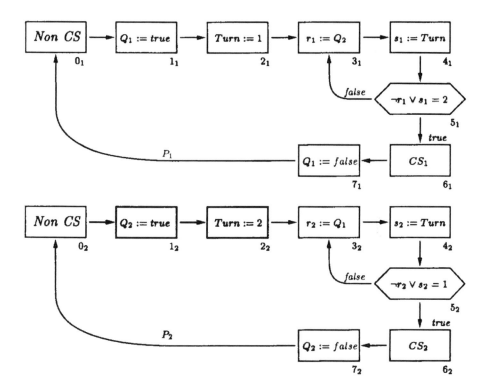

FIGURE 1.2. Mutual exclusion flowchart.

state is an initial state, and such that the transition from one state to the next is caused by an execution of one of the labeled instructions by P_1 or P_2. To explain this, we must first define states. For this, two "control variables" are added, CP_1 and CP_2. CP_1 (Control Position of P_1) takes values in $\{0_1, \ldots, 7_1\}$, and CP_2 takes values in $\{0_2, \ldots, 7_2\}$. A state is an assignment of a value to each of the variables

$$CP_1, CP_2, Q_1, Q_2, Turn, r_1, s_1, r_2, s_2$$

in its range, namely a function from the set of variables into their ranges. There are thus $8^2 \times 2^7$ states, which explains why the analysis of the protocol is difficult.

An initial state is a state σ such that $\sigma(CP_1) = 0_1$, $\sigma(CP_2) = 0_2$, $\sigma(Q_1) = false$, and $\sigma(Q_2) = false$. The other values of σ are immaterial. A "transition" is a pair of states (σ_1, σ_2) that represent an execution of one of the instructions. A transition thus corresponds to an arrow of the flowchart. To each arrow α in the flowchart of P_1 or P_2 we attach a set of transitions $Tr(\alpha)$. I give four representative examples:

(1) If $\alpha = (0_2, 1_2)$ then $Tr(\alpha)$ consists of the set of all pairs of states (σ_1, σ_2) such that $\sigma_1(CP_2) = 0_2$, $\sigma_2(CP_2) = 1_2$, and for any other variable v $\sigma_1(v) = \sigma_2(v)$. Similarly, the transitions corresponding to the arrow leading from 6_2 (the CS_2 node) to 7_2 change only the control variable of P_2.

(2) If $\alpha = (1_2, 2_2)$ then $Tr(\alpha)$ is the set of all pairs of states (σ_1, σ_2) such that $\sigma_1(CP_2) = 1_2$, $\sigma_2(CP_2) = 2_2$, $\sigma_2(Q_2) = true$, but all other variables do not change.

(3) If $\alpha = (3_2, 4_2)$, then $Tr(\alpha)$ is the set of all transitions (σ_1, σ_2) such that control CP_2 moves from 3_2 to 4_2, $\sigma_2(r_2) = \sigma_1(Q_1)$ and all other variables do not change. The effect of reading register Q_1 and assigning the value obtained to r_2 is thus expressed in the relation $\sigma_2(r_2) = \sigma_1(Q_1)$ holding between the state σ_1 and the resulting state σ_2.

(4) Finally, the transitions attached to the *true* and to the *false* edges are defined: for $\alpha = (5_2, 6_2)$, the *true* arrow, $Tr(\alpha)$ is the set of all transitions (σ_1, σ_2) such that $\sigma_1(CP_2) = 5_2$, $\sigma_1(r_2) = false \vee \sigma_1(s_2) = 1$, and $\sigma_2(CP_2) = 6_2$; all variables other than CP_2 do not change. The transitions attached to $(5_2, 3_2)$ are similarly defined, except that the condition $\neg r_2 \vee s_2 = 1$ must be evaluated to false at state σ_1.

In a similar vein transitions are attached to the edges of the flowchart of P_1.

Histories describe runs of the protocol: a *history* is an infinite sequence of states $\sigma_0, \sigma_1, \ldots$ such that σ_0 is an initial state and for each i either $\sigma_i = \sigma_{i+1}$ or (σ_i, σ_{i+1}) is a transition in some $Tr(\alpha)$ where α is an arrow either in the flowchart of P_1 or of P_2. I want the possibility of repeating the same state several times in order to allow indefinite stays at the external (zeroth) nodes and at the CS nodes.

The mutual-exclusion property is that in every history there is no state σ_i such that $\sigma_i(CP_1) = 6_1$ and $\sigma_i(CP_2) = 6_2$, namely such that both processes are in their critical sections. Since histories are not structures (interpretations) in the sense defined in this chapter, we cannot express the mutual-exclusion property as a first-order statement that holds there. We shall therefore change our point of view a little and transform histories into structures. This is achieved by viewing the state index i in σ_i as an individual of our universe, and the protocol variables as functions. Thus instead of $\sigma_i(CP_1) = 4_1$ we shall write $Val_{CP_1}(i) = 4_1$ to say that the control of the ith state is at 4_1, etc. In this mode, Val_{CP_1} is a function (the control position function) rather than a variable, and index i is its argument.

To make a precise definition, a signature L is defined first, consisting of the following:

(1) A binary relation symbol $<$. (The intended interpretation of $<$ is the standard ordering on the natural numbers taken as the universe.)

(2) A unary function symbol $Succ$ (the "successor" function). The intended interpretation of $Succ(i)$ is $i+1$. Since $Succ$ can be defined from $<$, $Succ$ is not strictly necessary, but we find it useful.

(3) For every protocol variable v a unary function Val_v is introduced. So there are nine unary function symbols in L which are called state functions: $Val_{CP_1}, Val_{CP_2}, Val_{Q_1}, Val_{Q_2}, Val_{Turn}, Val_{r_1}, Val_{s_1}, Val_{r_2}, Val_{s_2}$.

(4) The following constants: $0, 1, 2,$ $true,$ $false,$ $0_1, \ldots, 7_1,$ and $0_2, \ldots, 7_2$. (The intended interpretation of $false$ is 0, and of $true$ is 1; constants $0_1, \ldots, 7_1, 0_2, \ldots, 7_2$ are interpreted as distinct numbers. 1 and 2 serve as values of $Turn$.)

A standard interpretation of L is one in which the universe is the set of all natural numbers and $<$ is the natural ordering. Given a standard interpretation $\mathcal{H}$ of L, an infinite sequence $\langle \sigma_i \mid i \in \mathbb{N} \rangle$ is defined by the relation

$$\sigma_i(v) = Val_v(i)$$

where v is one of the nine variables $CP_1, CP_2, Q_1, Q_2, Turn, r_1, s_1, r_2, s_2$. σ_i may fail to be a state (if $Val_v(i)$ is not in the range of variable v), and the sequence $\langle \sigma_i \mid i \in \mathbb{N} \rangle$ may fail to be a history. If $\langle \sigma_i \mid i \in \mathbb{N} \rangle$ is a history (σ_0 is an initial state and for every i either $\sigma_i = \sigma_{i+1}$ or $\langle \sigma_i, \sigma_{i+1} \rangle$ is a transition) then $\mathcal{H}$ is said to be a *history model*. Equivalently, a history model is defined to be a standard interpretation of L that satisfies the following finite set T of first-order axioms (sentences). These sentences ensure that their model represents an execution of the protocol.

T1: Each of the functions gives values in its intended range. For example the part that deals with CP_1 is

$$\forall i (Val_{CP_1}(i) = 0_1 \vee Val_{CP_1}(i) = 1_1 \vee \cdots \vee Val_{CP_1}(i) = 7_1).$$

T2: All constants are distinct. That is $0_1 \neq 1_1 \wedge 0_1 \neq 2_1 \ldots$ etc.

T3: The first state, the zeroth state, is an initial state:

$$Val_{CP_1}(0) = 0_1 \wedge Val_{CP_2}(0) = 0_2 \wedge Val_{Q_1}(0) = false \wedge Val_{Q_2}(0) = false.$$

To complete the definition of T and to express the relationship between each state and its successor we need the following notation. If $F = \{f_1, \ldots, f_k\}$ is a

set of unary function symbols, then the formula $id_F(i, j)$ with free variables i, j
is the following:

$$(3) \qquad f_1(i) = f_1(j) \wedge f_2(i) = f_2(j) \wedge \ldots \wedge f_k(i) = f_k(j).$$

When F is omitted from id_F, then we mean that F is the set of all state functions.
Thus $id(i, j)$ says that the ith state is identical with the jth state.

For each arrow α in the flowchart for P_1 or for P_2 we define a formula $\tau_\alpha(i)$ with
a free variable i. $\tau_\alpha(i)$ says that the pair of states (σ_i, σ_{i+1}) forms a transition
in $Tr(\alpha)$. I give four representative examples.

(1) For $\alpha = (0_1, 1_1)$ the formula $\tau_{(0_1, 1_1)}(i)$ is the conjunction

$$Val_{CP_1}(i) = 0_1 \wedge Val_{CP_1}(Succ(i)) = 1_1 \wedge id_F(i, Succ(i))$$

where F is the collection of all state functions except Val_{CP_1}.

(2) $\tau_{(1_1, 2_1)}(i)$ is the formula

$$Val_{CP_1}(i) = 1_1 \wedge Val_{CP_1}(Succ(i)) = 2_1 \wedge$$
$$Val_{Q_1}(Succ(i)) = true \wedge id_F(i, Succ(i))$$

where F is the collection of all state functions except Val_{CP_1} and Val_{Q_1}.

(3) For $\alpha = (4_1, 5_1)$, $\tau_\alpha(i)$ is the formula

$$Val_{CP_1}(i) = 4_1 \wedge Val_{CP_1}(Succ(i)) = 5_1 \wedge$$
$$Val_{s_1}(Succ(i)) = Val_{Turn}(i) \wedge id_F(i, Succ(i))$$

where F is the collection of all state functions except Val_{CP_1} and Val_{s_1}.

(4) For $\alpha = (5_1, 6_1)$, $\tau_\alpha(i)$ is $Val_{CP_1}(i) = 5_1 \wedge Val_{CP_1}(Succ(i)) = 6_1 \wedge$
$(Val_{r_1}(i) = false \vee Val_{s_1}(i) = 2) \wedge id_F(i, Succ(i))$, where F is the set
of all state functions except Val_{CP_1}.

The last axiom T4 can now be described. It is a long disjunction over the set
$ARROWS$ of all arrows in P_1 and P_2.

$$\forall i \ [id(i, Succ(i)) \vee \bigvee\nolimits_{\alpha \in ARROWS} \tau_\alpha(i)].$$

A standard model of T is called a history model, that is, a model which augments
$(\mathbf{N}, +, <)$ with an interpretation of L. The mutual exclusion property MUTEX is
the following sentence:

$$\forall i \neg (Val_{CP_1}(i) = 6_1 \wedge Val_{CP_2}(i) = 6_2).$$

We are going to show that $T \Longrightarrow$ MUTEX. That is, in every history model MUTEX
holds. There are other properties of this protocol that one may want to prove.
For example the liveness property says that each process can access its critical
section if there are sufficiently many transitions by this process and if the other
process has only finite stays in its critical sections. (A detailed analysis is left to
the reader.)

3.1. Proof of the Mutual Exclusion property. A possible approach to proving the mutual exclusion property is to find out all states that can appear in any history of the protocol (by computing the minimal set (under inclusion) containing all initial states and closed under each of the protocol's transitions) and then to check that none of these states violates the mutual exclusion property. While correct, this solution is not satisfactory because of the enormous work it requires, and since it is not sufficiently general. We are interested in solutions that also give a certain explanation of why the protocol works, and we turn to a different approach.

Let $\mathcal{H}$ be a model history (as defined above) and we shall prove that MUTEX holds in $\mathcal{H}$. For this we shall state and prove a series of lemmas.

For easier expression, let us agree in the following to write $Val_{CP_1} = 0_1, 1_1$, for example, instead of $Val_{CP_1}(i) = 0_1 \vee Val_{CP_1}(i) = 1_1$. Also, universal quantifiers in front of formulas can be omitted. So, for example, (2) below is a shorthand for $\forall i \in \mathbf{N}(Val_{CP_1}(i) = 0_1 \vee Val_{CP_1}(i) = 1_1 \longrightarrow Val_{Q_1}(i) = false)$.

LEMMA 3.1.   (1) $Val_{CP_1} = 2_1, 3_1, 4_1, 5_1, 6_1, 7_1 \longrightarrow Val_{Q_1} = true$.
(2) $Val_{CP_1} = 0_1, 1_1 \longrightarrow Val_{Q_1} = false$.
(3) $Val_{CP_2} = 2_2, 3_2, 4_2, 5_2, 6_2, 7_2 \longrightarrow Val_{Q_2} = true$.
(4) $Val_{CP_2} = 0_2, 1_2 \longrightarrow Val_{Q_2} = false$.

Proof. We prove the first two items and argue by symmetry for the last two. The antecedent of (1) $Val_{CP_1} = 2_1, 3_1, 4_1, 5_1, 6_1, 7_1$ is a shorthand for the formula $Val_{CP_1}(i) = 2_1 \vee Val_{CP_1}(i) = 3_1, \ldots, \vee Val_{CP_1}(i) = 7_1$, which we denote by $\phi(i)$. The consequence, denoted $\psi(i)$ is the formula $Val_{Q_1}(i) = true$, so that the first sentence is in fact $\forall i(\phi(i) \longrightarrow \psi(i))$. We prove by induction on i that $\phi(i) \longrightarrow \psi(i)$ holds, and that the second item $Val_{CP_1}(i) = 0_1, 1_1 \longrightarrow Val_{Q_1}(i) = false$ holds too.

For $i = 0$, $Val_{CP_1}(0) = 0_1$ by T3 (which is the third sentence in T) and so the antecedent $Val_{CP_1} = 2_1, 3_1, 4_1, 5_1, 6_1, 7_1$ does not hold (by T2 which says that 0_1 is distinct from the other locations) and the first sentence holds in this case. The second sentence is true by T3 which says that $Val_{Q_1}(0) = false$.

Now assume the inductive hypothesis for i and we shall prove it for $i' = i + 1$. We know that either $id(i, i')$ or $\tau_\alpha(i)$ holds for some α, and we must check all possibilities for α. If $id(i, i')$ or $\tau_\alpha(i)$ holds with α an arrow in the flowchart for P_2, then $Val_{CP_1}(i) = Val_{CP_1}(i')$ and $Val_{Q_1}(i) = Val_{Q_1}(i')$ and hence stage i' follows from the inductive assumption. So now assume α is in the flowchart of P_1. We must check one after the other all possibilities. For example, if $\alpha = (7_1, 0_1)$ and $\tau_\alpha(i)$, then necessarily $Val_{CP_1}(i') = 0_1$ and $Val_{Q_1}(i') = false$. It can be seen that this implies the truth of the inductive hypothesis for i'. $\square$

LEMMA 3.2.   (1) $(Val_{CP_1} = 5_1, 6_1, 7_1) \wedge Val_{s_1} = 2 \longrightarrow Val_{Turn} = 2$.
(2) $(Val_{CP_2} = 5_2, 6_2, 7_2) \wedge Val_{s_2} = 1 \longrightarrow Val_{Turn} = 1$.

Proof. The proof is by induction, and we present the key points informally. Again we look only at (1) and argue that (2) is obtained similarly.

Suppose that $i' = i + 1$ and that $(Val_{CP_1}(i') = 5_1, 6_1, 7_1) \wedge Val_{s_1}(i') = 2$. We must prove that $Val_{Turn}(i') = 2$. If $id(i, i')$ then this is clear by the inductive assumption since $\sigma_i = \sigma_{i+1}$ in this case. Otherwise $\tau_\alpha(i)$ holds for some α

in P_1 or in P_2. Suppose first α is in P_1. Assume $Val_{CP_1}(i') = 5_1$. Then $Val_{CP_1}(i) = 4_1$ follows, and, as $Val_{s_1}(i') = 2$, $Val_{Turn} = 2$ as well. (Informally this is clear from the way the transition for $(4_1, 5_1)$ is defined, and more formally this should be a consequence of $\tau_{(4_1,5_1)}(i)$.) To continue the proof, assume that $Val_{CP_1}(i') = 6_1, 7_1$. Then $Val_{CP_1}(i) = 5_1, 6_1$ and induction can be applied to the ith state to yield that $Val_{Turn}(i) = 2$. We want to conclude that $Val_{Turn}(i') = 2$ as well. Observe that P_1 can falsify the statement $Val_{Turn} = 2$ only by writing 1 on $Turn$ in executing node 2_1, but this is not the case.

Now suppose that α is in P_2. An inspection of all the formulas $\tau_\alpha(i)$ for α in P_2 shows that $Val_{CP_1}(i) = Val_{CP_1}(i')$ and $Val_{s_1}(i) = Val_{s_1}(i')$ holds, and also that if $Val_{Turn}(i) \neq Val_{Turn}(i')$ then necessarily $Val_{Turn}(i') = 2$ (P_2 writes only 2 on $Turn$). So, since (1) holds for i, it remains true for i'.

In the next lemma it is convenient to write CS_1 for node 6_1 and CS_2 for 6_2.

LEMMA 3.3.   (1) $Val_{CP_1} = CS_1 \longrightarrow Val_{r_1} = false \vee Val_{s_1} = 2$.
(2) $Val_{CP_2} = CS_2 \longrightarrow Val_{r_2} = false \vee Val_{s_2} = 1$.

The proof is left to the reader. The main lemma needed for the mutual exclusion is presented now.

LEMMA 3.4. Let u, v and w be the following formulas:

$$u \ = \ (Val_{Q_2} = false),$$
$$v \ = \ (Val_{CP_2} = 2_2),$$
$$w \ = \ (Val_{CP_2} = 3_2, 4_2, 5_2) \wedge$$
$$Val_{Turn} = 2 \wedge$$
$$(Val_{CP_2} = 5_2 \longrightarrow Val_{s_2} = 2) \wedge$$
$$(Val_{CP_2} = 4_2, 5_2 \longrightarrow Val_{r_2} = true).$$

Then

$$(Val_{CP_1} = 4_1, 5_1, 6_1) \wedge Val_{r_1} = false \longrightarrow u \vee v \vee w.$$

A symmetric statement is obtained by exchanging indices 1 and 2.

The proof is again by induction on i, and by checking all possible arrows. I guide the reader only through the main points of the argument.

Assume that $i' = Succ(i)$ and that the claim holds for i. If $id(i, i')$ then obviously the claim holds for i' as required. We shall consider two cases: that the move related to (i, i') is by P_1 (case 1) and by P_2 (case 2).

Case 1: (i, i') is a P_1 step. There are three sub-cases here according as to which is the value of $Val_{CP_1}(i')$: $4_1, 5_1, 6_1$.

4_1: Suppose that $Val_{CP_1}(i') = 4_1$ and $Val_{r_1}(i') = false$, and we prove $u(i') \vee v(i') \vee w(i')$. Since arrow $(3_1, 4_1)$ was executed in step (i, i'), $Val_{r_1}(i') = Val_{Q_2}(i) = Val_{Q_2}(i')$ and thence $Val_{Q_2}(i') = false$ which is $u(i')$. Hence the disjunction holds.

5_1: Suppose that $Val_{CP_1}(i') = 5_1$ and $Val_{r_1}(i') = false$. Then $Val_{CP_1}(i) = 4_1$ and $Val_{r_1}(i) = false$ and hence $u(i) \vee v(i) \vee w(i)$ holds by the inductive assumption. Since step (i, i') is done by P_1 in executing $(4_1, 5_1)$ which never changes the value of register $Turn$, and as u, v, w speak only about local variables that may be changed by P_2, $u(i') \vee v(i') \vee w(i')$ holds as well.

6_1: A similar argument works for the case that $Val_{CP_1} = 6_1$.

Case 2: (i, i') is a P_2 step. In this case the variables of P_1 do not change. Assume $Val_{CP_1}(i') = 4_1, 5_1, 6_1$, and $Val_{r_1}(i') = false$. We must prove $(u \lor v \lor w)(i')$. Since $Val_{CP_1}(i') = Val_{CP_1}(i)$, $Val_{CP_1}(i) = 4_1, 5_1, 6_1$ and similarly $Val_{r_1}(i) = false$. So $u(i) \lor v(i) \lor w(i)$ holds by the inductive assumption. In order to prove that $u(i') \lor v(i') \lor w(i')$, we consider three sub-cases:

(1) Assume that $u(i)$ holds. That is $Val_{Q_2}(i) = false$. Then either $Val_{Q_2}(i') = false$ as well (which ends this sub-case) or else the step (i, i') corresponds to $(1_2, 2_2)$ (which is the only possibility for a transition that writes *true* on Q_2). Hence $Val_{CP_2}(i') = 2_2$ (if the second possibility happens) which shows $v(i')$.

(2) Assume that $v(i)$ holds, that is $Val_{CP_2}(i) = 2_2$. Hence $Val_{CP_2}(i') = 3_2$ and $Val_{Turn}(i') = 2$. This clearly implies $w(i')$.

(3) Assume $w(i)$ and then we derive $w(i')$. $w(i)$ implies $Val_{CP_2}(i) = 3_2, 4_2, 5_2$. To these three possibilities correspond (respectively) possibilities

$$Val_{CP_2}(i') = 4_2, 5_2, 3_2.$$

Why did we exclude the possibility of step $(5_2, 6_2)$ for P_2? Because if $Val_{CP_2}(i) = 5_2$, then $w(i)$ implies that the condition for taking the *true* arrow does not hold. This proves the first conjunct for $w(i')$.

Now $Val_{Turn}(i') = 2$ is an obvious consequence of $Val_{Turn}(i) = 2$ and the fact that only P_1 can change $Turn$ to 1.

To prove the third conjunct of $w(i')$ assume that $Val_{CP_2}(i') = 5_2$ and we shall see that $Val_{s_2}(i') = 2$. Well, this follows since $\tau_{(4_2, 5_2)}$ implies that $Val_{s_2}(i') = Val_{Turn}(i)$, and since $Val_{Turn}(i) = 2$ by $w(i)$.

Finally we prove the fourth conjunct of $w(i')$. Assume that $Val_{CP_2}(i') = 4_2, 5_2$ and we shall derive $Val_{r_2}(i') = true$. If $Val_{CP_2}(i') = 4_2$, then $Val_{r_2}(i') = Val_{Q_1}(i) = true$ (this last equality is a consequence of the fact that $Val_{CP_1}(i) = 4_1, 5_1, 6_1$ and Lemma 3.1(1)). If $Val_{CP_2}(i') = 5_2$ then $Val_{CP_2}(i) = 4_2$, and the fact that $w(i)$ is assumed implies that $Val_{r_2}(i) = true$ (by the fourth conjunct) and hence that $Val_{r_2}(i') = true$ (since Val_{r_2} is not changed). $\square$

At long last we shall obtain the mutual exclusion sentence MUTEX. So assume that for some i both $Val_{CP_1}(i) = CS_1$ and $Val_{CP_2}(i) = CS_2$ hold, and we shall find a contradiction.

By Lemma 3.3, both

(4) $\quad Val_{r_1}(i) = false \lor Val_{s_1}(i) = 2$ and $Val_{r_2}(i) = false \lor Val_{s_2}(i) = 1$

hold. In case $Val_{r_1}(i) = false$ (a similar argument can be given in case $Val_{r_2}(i) = false$) Lemma 3.4 implies $u(i) \lor v(i) \lor w(i)$. $u(i)$ is $Val_{Q_2}(i) = false$ and this contradicts Lemma 3.1(3) as $Val_{CP_2}(i) = 6_2$. $v(i)$ is $Val_{CP_2}(i) = 2_2$ which clearly contradicts $Val_{CP_2}(i) = 6_2$. So $w(i)$ remains, which implies $Val_{CP_2}(i) = 3_2, 4_2, 5_2$, and this is again a contradiction.

So now assume $Val_{r_1}(i) = true$ and $Val_{r_2}(i) = true$. Hence $Val_{s_1}(i) = 2$ and $Val_{s_2}(i) = 1$ (by formula (4)). Then by Lemma 3.2(1)(2), both $Val_{Turn}(i) = 2$ and $Val_{Turn}(i) = 1$!

2
System executions

Time is a fundamental issue in concurrency, and we begin this chapter with some observations that date back to Bertrand Russell and Norbert Wiener concerning "interval orderings" and time. Then we investigate finiteness conditions, define system executions (adapting Lamport's original definition), and discuss higher and lower-level events. In the last section we analyze the paradox of Achilles and the Tortoise and its relation with the finiteness condition.

1. Time and interval orderings

Modeling is not only an abstract description of reality, but also a simplification which disregards some non-essential features. For many applications in computer science the occurrence time of each event is not needed, and what really counts is the *precedence* relation. This relation is a partial ordering which holds for events a and b if a ends before b begins. So two events are incomparable if their temporal extensions overlap. Two simple and useful properties of this precedence relation reflect its nature: The Russell–Wiener property, and the finiteness property of Lamport.

We begin with the definition of interval orderings.

DEFINITION 1.1. *(Interval orderings) Let $(L, <_L)$ be a linear ordering and S be a set of non-empty intervals of L. (An interval is a convex subset of L, i.e. a set $I \subseteq L$ such that if $a, b \in I$ then $a <_L x <_L b$ implies $x \in I$.) Then define for $I, J \in S$, $I \prec_L J$ iff $\forall x \in I \, \forall y \in J (x <_L y)$. We say that I precedes J when $I \prec_L J$.*

It is easy to prove that $\prec_L$ is a partial ordering of S, that is to say a transitive and irreflexive relation.

The orderings obtained by means of Definition 1.1 above are called *interval orderings* . (Fishburn [16] is devoted to this subject.)

24

More formally, the following definition says that an interval ordering is any partial ordering that is *isomorphic* to some $\prec_L$ ordering of a set S of intervals of some linear ordering $(L, <_L)$.

DEFINITION 1.2. *Let R be a partial ordering of a set E. A representation of (E, R) is a linear ordering $(L, <_L)$ and a map μ of E into intervals of L such that for all $a, b \in E$, aRb iff $\mu(a) \prec_L \mu(b)$.*

An interval ordering is defined to be an ordering that has a representation.

For concreteness, the intervals in a representation may be assumed to be closed to the left and open to the right. That is, of the form $I = [a, b)$ where a is denoted *Left_End(I)*, and b is denoted *Right_End(I)*. Then we have $I \prec_L J$ iff *Right_End(I)* $\leq_L$ *Left_End(J)*. This convention is not essential, but it has the advantage that if two intervals have a non-empty intersection then their intersection is again an interval of this form. (A single point is certainly not of this form; so if we want point intervals we must admit closed intervals as well.)

Two intervals I and J are said to overlap iff $I \cap J \neq \emptyset$. Since L is a linear ordering, for any intervals I and J either I and J are comparable ($I \prec_L J$ or $J \prec_L I$) or else I and J overlap. Thus overlapping coincides with incomparability.

The following property of $\prec_L$ is of prime importance. Suppose that $A \prec_L B$ and $C \prec_L D$ are intervals such that B and C overlap; then $A \prec_L D$. To prove this pick any point $p \in B \cap C$. Then $\forall a \in A$ ($a <_L p$) follows from $A \prec_L B$ since $p \in B$. Similarly $\forall d \in D$ ($p <_L d$) follows. Hence, for any $a \in A$ and $d \in D$, $a <_L p <_L d$, and thus $a <_L d$. This means that $A \prec_L D$. Bertrand Russell and Norbert Wiener [31] proved that this property characterizes interval orderings, as we are going to show next (see also Fishburn [16] and Lamport [20].) In fact, instead of requiring that B and C overlap, we may simply ask that $\neg(C \prec_L B)$. It turns out that the relation $\neg(C \prec_L B)$ is so significant that we need a special symbol to denote it. So we define

$$x \not\succ_L y \text{ iff } \neg(y \prec_L x).$$

With this symbol, the property of $\prec_L$ discussed above is:

$$A \prec_L B \not\succ_L C \prec_L D \longrightarrow A \prec_L D.$$

This is easier to remember because of its resemblance with transitivity.

DEFINITION 1.3 (THE RUSSELL–WIENER PROPERTY). *Let (E, R) be a partial-ordering. The Russell–Wiener property is the following:*

(5) $$\forall a, b, c, d \in E(aRb \wedge \neg cRb \wedge cRd \longrightarrow aRd)$$

We have seen for any linear ordering $(L, <)$ and a set of intervals of L that the Russell–Wiener property holds for the relation $\prec_L$.

THEOREM 1.4 (RUSSELL AND WIENER). *A partial ordering satisfies the Russell–Wiener property (5) iff it has a representation.*

Proof. [Wiener [31]] We have already argued that an interval ordering $\prec_L$ satisfies the Russell–Wiener property.

Assume now an ordering (E, R) that satisfies the Russell–Wiener property, and we shall find a representation. We say that $X \subseteq E$ is an *antichain* iff it is a set of pairwise incomparable elements: $\forall x \neq y \in X(\neg xRy \text{ and } \neg yRx)$.

A maximal antichain is called a *moment* (it may be finite or infinite). An antichain X is thus a moment iff any y not in X is comparable with some $x \in X$.

Let L be the collection of all moments, and define for $X, Y \in L$,

$$X < Y \text{ iff } \exists x \in X, \exists y \in Y \ (xRy).$$

We prove first that $<$ is a linear order on L. Transitivity of $<$ is an easy consequence of the Russell–Wiener formula: If $X < Y < Z$ then there are $x_0 \in X$ and $y_0 \in Y$ such that $x_0 R y_0$ (as $X < Y$), and similarly there exists $y_1 \in Y$ and $z_1 \in Z$ such that $y_1 R z_1$. But as Y is an antichain, y_0 and y_1 are equal or incomparable. So $x_0 R z_1$ follows the Russell–Wiener property. Hence $X < Z$.

Clearly $<$ is irreflexive. Linearity follows from the maximality of moments: if X and Y are two distinct maximal antichains then none is included in the other. Take some $x \in X \setminus Y$ and then x is comparable with some y in Y. This implies that X and Y are $<$-comparable.

Now we define the representation map μ from E into intervals of L. Let $\mu(e)$ for $e \in E$ be the collection of all moments X such that $e \in X$. Since any $e \in E$ is contained in at least one maximal antichain, the representing set $\mu(e)$ is nonempty. (It seems that we need here the axiom of choice, but in applications finite antichains suffice because the finiteness condition implies that all antichains are finite.)

We claim that

(1) $\mu(e)$ is convex (an interval in L), and
(2) for all $e, f \in E$, eRf iff $\mu(e) \prec_L \mu(f)$.

To prove the convexity, assume $X < Y < Z$, and $X, Z \in \mu(e)$. Then $e \in X$ and $e \in Z$. We will prove that $Y \in \mu(e)$. Assume for the sake of a contradiction that $e \notin Y$. By maximality of Y, there is $y \in Y$ such that eRy or yRe. Assume for example that eRy. Since $Y < Z$ there are $y' \in Y$ and $z' \in Z$ such that $y'Rz'$. Then the Russell–Wiener formula applies to e, y, y', z' and implies that eRz', but this contradicts the fact that the moment Z is an antichain.

In order to prove the second part of the claim, assume first that eRf. Suppose $X \in \mu(e)$ and $Y \in \mu(f)$. Then $e \in X$ and $f \in Y$ by definition of μ, and so $X < Y$ by definition of $<$. Thus $\mu(e) \prec_L \mu(f)$.

Next we check that if $\neg eRf$ then $\neg(\mu(e) \prec_L \mu(f))$. Assume $\neg eRf$. If fRe then $\mu(f) \prec_L \mu(e)$, as proved above, and hence $\neg(\mu(e) \prec_L \mu(f))$ by the transitivity and irreflexivity of $\prec_L$. If e and f are incomparable under R then a maximal antichain exists which includes both e and f. And again this shows $\neg(\mu(e) \prec_L \mu(f))$. □

This proof is the one given by Wiener [31]; it follows Russell's intuition which wants to base the concept of a temporal point (moment) on the more elementary notions of event and precedence. The proof in Fishburn [16] is more constructive.

1.1. Finiteness conditions. I think no one will contest the assumption that any system can only contain finitely many event occurrences that precede a given instant. In fact, since the number of concurrently operating agents is finite, only finitely many events can even begin before any given instant t, and hence all others begin after t. Leslie Lamport [20] proposed and showed the applicability of the following property.

DEFINITION 1.5 (FINITENESS PROPERTY). *A partial ordering $(E, <)$ is said to satisfy the finiteness condition (of Lamport) iff for every $e \in E$ the set $\{x \in E \mid \neg(e < x)\}$ is finite.*

A weaker property is the "finite predecessors property". A partial ordering $(E, <)$ satisfies the finite predecessors property iff for every $e \in E$ the set $\{x \in E \mid x < e\}$ is finite.

Observe that Lamport's finiteness condition implies the finite predecessors property and it also implies that every antichain is finite.

The finite predecessors property enables induction. To prove that $\forall x \phi(x)$ holds in $(E, <)$ we assume towards a contradiction that for some $e \in E$ $\phi[x/e]$ does not hold, and take e to be minimal with this property. That is, for any $e' < e$, $\phi[x/e']$ does hold. If we can derive a contradiction from this situation then $\forall x \phi$ is proved. (The notation $\phi[x/e]$ was introduced on page 11.) Somewhat more complex forms of induction will be described below.

In fact, Lamport distinguishes between terminating and nonterminating events. Intuitively, terminating events have a bounded (finite) duration, while nonterminating events endure forever. A nonterminating event can never be followed by another event. Formally we have

DEFINITION 1.6 (LAMPORT). *Global-time models: Let S be a structure for the signature $\{E, \prec, T\}$ where E is the unique sort (the universe), $\prec$ is a binary predicate, and T a unary predicate (we shall often write Terminating, in full, instead of T). Members of E are called events, T is a subset of E called the set of terminating events, and $\prec$ is called the precedence relation. Suppose that $\vdash$ is defined by a $\vdash b$ iff $\neg(b \prec a)$. Then S is called a* global-time model *iff:*

(1) $\forall a, b, c, d \ (a \prec b \vdash c \prec d \longrightarrow a \prec d)$. *(This is the Russell–Wiener property. So $\prec^S$ is an interval ordering.)*

(2) *Every antichain is finite, and the finite predecessors property holds: that is for any $a \in E$ the set $\{x \in E \mid x \prec a\}$ is finite.*

(3) *If $\neg T(e)$ then there is no $x \in E$ such that $e \prec x$. In other words, if an event e is followed by some event x, then e must be terminating.*

(4) *For any a such that $T(a)$, $\{x \in E \mid x \vdash a\}$ is finite. (The finiteness property of Lamport, for terminating events only: the set of events that follow a terminating event is co-finite[1].)*

---

[1] A subset is co-finite if its complement is finite

Intuitively, the finiteness property of Lamport says that if a is terminating, then most of the events follow a: "most" in the sense that there is only a finite number of exceptions. So this property is meaningful (non trivial) only if the set of events is infinite. To the reader who is familiar with the notations of Lamport in [20] I should say that $\prec$ here corresponds to $\longrightarrow$ and $\vdash$ corresponds to $\cdots\cdots\rightarrow$.

Claim: It follows from (2) and (3) that the nonterminating events form a finite antichain, and as (4) implies that there can be only countably many terminating events it follows that a global-time model is countable.

Proof of the claim that the number of events in a global-time model is countable. For any terminating event e in E let $p(e)$ be the number of events x with $x \prec e$. The finite predecessor property implies that $p(e)$ is finite, and we define for every $n \in \mathbb{N}$ $E_n = \{e \in E \mid T(e) \wedge p(e) = n\}$. Then $\bigcup_{n\in\mathbb{N}} E_n$ is the set of all terminating events and it suffices to prove that every E_n is finite to conclude that the number of all terminating events is countable, and hence that the number of all events is countable (as the nonterminating events form an antichain which is finite). If some E_n were infinite, pick $a \in E_n$ and apply the finiteness condition (4) to obtain that $X = \{x \in E \mid x \not\vdash a\}$ is finite. So there is some $e \in E_n \setminus X$ since E_n is infinite. But then $e \succ a$ and hence all predecessors of a—and a itself—are predecessors of e, which shows that $p(e) \geq n + 1$, in contradiction to $e \in E_n$.

If S is a global-time model then S has the form $S = (E^S, \prec_S, T^S)$ where E^S, $\prec_S$, T^S are the interpretations of the signature symbols. However we often write $S = (E, \prec, T)$ or $(E, \prec_S, T)$ for notational simplicity.

DEFINITION 1.7 (REPRESENTATIONS). *Let $(L, <_L)$ be a linear ordering and $S = (E, <_S, T)$ be a global-time model. We say that μ is a representation of S in L iff μ is a representation as in Definition 1.2 in which terminating events are represented by bounded intervals, and nonterminating events are represented by right unbounded intervals. That is:*

(1) *For each $e \in E$, $\mu(e)$ is an interval in L, and $e_1 <_S e_2$ iff $\mu(e_1) \prec_L \mu(e_2)$.*
(2) *If $T(e)$ holds, then $\mu(e)$ is of the form $[a, b)$, but if $\neg T(e)$ holds (that is e is nonterminating) then $\mu(e)$ has the form $[a, \infty) = \{x \in L \mid a \leq_L x\}$.*

Theorem 1.4 implies that any global-time model has a representation as intervals of natural numbers. That is, $(L, <_L)$ can be chosen as $(\mathbb{N}, <)$. To see this, consider a representation μ of $(E, <_S)$ into any linear ordering $(L, <_L)$. Then make a list $\langle e_n \mid n \in \mathbb{N}\rangle$ of all terminating events in E (assuming E is infinite, for otherwise the claim is obvious). Define by induction a sequence a_k of points in L. a_0 is any point in $\mu(e_0)$. At the nth stage of the construction consider e_n and the finite set of points $P_n = \{a_0 \dots\}$ constructed so far. Define P_{n+1} in such a way that $P_{n+1} \cap \mu(e_n) \neq \emptyset$, and for every $m < n$ $\mu(e_n) \cap \mu(e_m) \neq \emptyset$, then $P_{n+1} \cap \mu(e_n) \cap \mu(e_m) \neq \emptyset$. That is P_{n+1} adds to P_n points as necessary for this evidence. Now the points in $P = \bigcup_{i\in\mathbb{N}}$ have order-type ω (that of $\mathbb{N}$), because P satisfies the finiteness property in L. Indeed, any a_k is in some $\mu(e_n)$, and there is some N such that for every $p \geq N$, $e_n <_S e_p$ and hence $\mu(e_n) \prec_L \mu(e_p)$. Thus the points added at the pth stage (for $p > N$) are greater than a_k (being in $\mu(e_p)$).

Consider the function $\mu'(e) = \mu(e) \cap P$ for $e \in E$. We claim that it is a representation such that $\mu'(e)$ is finite iff e is terminating. Clearly every $\mu'(e)$ is nonempty. The finiteness of $\mu'(e)$ for terminating events was argued already when e_k was considered. If e is nonterminating, then for some K for all $k \geq K$, $\mu(a_k)$ is a subinterval of $\mu(e)$. (Since there are only finitely many events e_1 such that $e_1 <_S e$, there is some e_1 which does not precede e. Now from some index on, each e_m satisfies $e_1 <_{\mathcal{M}} e_m$, and then $\mu(e_m)$ is a subinterval of $\mu(e)$ as required.) This shows that if e is nonterminating then $\mu'(e)$ is infinite.

Clearly $e_1 <_S e_2$ implies $\mu'(e_1) \prec_P \mu'(e_1)$, but the other direction is also true, by construction. That is, if e_1 and e_2 are incomparable, then $\mu(e_1)$ and $\mu(e_2)$ intersect, and hence $\mu'(e_1)$ and $\mu'(e_2)$ intersect.

The finiteness properties of Lamport enable induction "on the right-end points" for the terminating events as explained here. Let $S = (E, <_S, T)$ be a global-time model. We have seen that there is a representation of S as intervals in $\mathbb{N}$, and fixing such a representation μ it is possible to prove by induction on n statements of the form: For every interval I in the range of μ, if n is the right-end point of the interval then $\phi(I)$ holds. It is neater to make such an inductive proof without referring to any representation.

Define for every terminating event e a set $P(e) = \{x \in E \mid \neg e <_S x\}$. The finiteness property of Lamport says that $P(e)$ is finite, and hence proofs by induction on $|P(e)|$ can be carried out. That is, to prove a statement of the form: "For all terminating events e $\psi(e)$ holds" argue as follows. Assume that event e is a counterexample with least value of $|P(e)|$, and derive a contradiction by finding another counterexample e_0 such that for some event u

$$e_0 <_S u, \text{ and } \neg(e <_S u).$$

We prove that $P(e_0) \subseteq P(e)$. Suppose $x \in P(e_0)$ but $x \notin P(e)$. Then $e <_S x$, and the Russell–Wiener property applies to

$$e_0 < u, \neg(e < u), e < x$$

and gives $e_0 <_S x$, in contradiction to $x \in P(e_0)$. (Theorem 4.7 in Chapter 4 is an example of such an inductive proof.) The following variant form of induction is quite useful.

For this aim, define a relation $e_1 R e_2$ on the terminating events by $e_1 R e_2$ iff for some event x $e_1 <_S x$ and $\neg(e_2 <_S x)$. If e_1, e_2 were intervals of some linear ordering, then $e_1 R e_2$ would correspond to $Right\_End(e_1) < Right\_End(e_2)$. Anyhow, it follows from the Russell–Wiener property that R is irreflexive and transitive. Define for any terminating event e $P(e) = \{d \in E \mid d R e\}$. Then the finiteness property implies that $P(e)$ is finite, because $P(e) \subseteq \{d \in E \mid d \not\vdash e\}$. Also, $e_1 R e_2$ implies $|P(e_1)| < |P(e_2)|$ (where $|X|$ is the cardinality of X). This enables induction on $|P(e)|$ which we call "induction on right-end(e)". To prove that $\phi(e)$ holds for every terminating event e, it suffices to prove that $\phi(e)$ follows whenever e is an event such that $\phi(d)$ holds for every event d with $d R e$.

2. Definition of system executions

A global-time model reflects only the temporal order type of the system run and no other information. Reality needs more to be described, and a richer language is necessary. A comprehensive model of executions of a program for example must give all the information that is relevant to the program's correctness. System executions are such models.

DEFINITION 2.1 (SYSTEM-EXECUTIONS AND SYSTEMS). *A structure S for a signature L is a system-execution iff*

(1) *L contains the signature $L_0 = \{E, \prec, T\}$, and it may contain other sorts and symbols besides those of L_0.*
(2) *The reduct of S to L_0 is a global-time model.*

Explicitly, a system-execution is a multi-sorted structure S containing a sort E^S of "events" on which there is a relation $\prec^S$ and a subset $T^S \subseteq E^S$ such that $(E^S, \prec^S, T^S)$ is a global-time model.

A System is a collection of system executions in some fixed signature L.

Remarks:

(1) A system-execution is a description of some part of reality; it is not a program or a prescription of how things should be. It is rather a model of some real or imaginary happening recorded by an omnipresent and all knowing agent. The relation $\prec$ is not necessarily the causal relation but the temporal precedence relation (which seems to me to be more objective).

(2) System-executions belong to the family of "event based" models for concurrency which emphasize events rather than states. It is different however from the event structures described by G. Winskel [32] mainly in the following point. An event structure describes the manifold of all possible realities—any event that could have possibly occurred is in the structure—and if two events are prohibited from occurring in one world then they are denoted as conflicting. A system execution describes a single world (or part of a world). It takes a collection of system executions in some given signature to describe and specify the manifold of all possibilities. Following Lamport [20], a collection of system executions is called a system.

(3) The term system execution was coined by Lamport [20] . I use this term to express my indebtedness and since I believe that my changes are in accordance with the original intuitions. A short description of Lamport's definition and a discussion of the global-time problem is given in the following.

2.1. Global time. This is a very short section (almost an extended footnote) which aims to give some definitions concerning the question of global time. The reader may want to pursue this issue with the papers cited at the end of the section.

Basing their arguments on modern physics, philosophers and scientists have offered alternative views to the temporal picture of a single, linearly ordered time axis that is adopted here. Lamport's analysis of distributed systems ([20]) is based on two precedence relations on operation executions $A \longrightarrow B$ for "precedence", and $A \dashrightarrow B$ which can be read as "A can affect B".

Lamport's Axioms for $\longrightarrow$ and $\dashrightarrow$ precedence relations.

(1) $\longrightarrow$ is irreflexive and transitive.
(2) $a \longrightarrow b$ implies $a \dashrightarrow b$ and $\neg b \dashrightarrow a$.
(3) $a \longrightarrow b \dashrightarrow c$ implies $a \dashrightarrow c$. And similarly $a \dashrightarrow b \longrightarrow c$ implies $a \dashrightarrow c$.
(4) $a \longrightarrow b \dashrightarrow c \longrightarrow d$ implies $a \longrightarrow d$.

The following structures were defined in [20]:

DEFINITION 2.2. *We say that $(E, \longrightarrow, \dashrightarrow, T)$ is a Lamport's structure iff it satisfies the four axioms listed above and moreover it satisfies the finiteness property: for every terminating $x \in E$ (i.e., x's such that $T(x)$) the following set is finite.*

$$\{y \in E \mid \neg x \longrightarrow y\}$$

Lamport discusses (and rejects) the *global-time* axiom.
Global-Time Axiom

For every a and b, $a \longrightarrow b$ iff $\neg b \dashrightarrow a$.

The global-time axiom is not a consequence of Lamport's four axioms, and if it is added then the temporal picture is much simplified. I prefer this simpler setting both for describing semantics of protocols and for investigating concurrency. Simplicity is a strong reason, but there are other reasons described in the papers by Ben-David[9]; Abraham, Ben-David and Magidor [5]; and Abraham, Ben-David and Moran [6]. The essence of these investigations is that, for a large family of protocols, correctness under the global-time assumption implies correctness under the more general axioms of Lamport.

An interesting investigation of Lamport's structures is in Anger [8].

2.2. Parallel composition of systems. We are going to define an operation $\parallel$ which describes the "parallel composition" of systems. The usefulness of this definition will be revealed only in Chapter 3, so the reader may well skip this subsection until it is needed. We begin with intersection and union of signatures.

DEFINITION 2.3. *Suppose that L_1 and L_2 are two signatures (or languages). We say that L_1 and L_2 agree on their common symbols if and only if:*

(1) *If F is any function symbol in both L_1 and L_2, then the arity of F in L_1 is the same as its arity in L_2, and the domain and range of F are the same sorts in L_1 and in L_2.*

(2) *If P is a predicate in L_1 and in L_2, then it has the same arity and is defined over the same sorts in L_1 and in L_2. For example if P is a predicate in both L_1 and L_2 that is unary in L_1, then it is unary in L_2 (and vice versa) and if it is defined on sort X in L_1, then X is a sort in L_2 as well and P is a subset of X in L_2.*

(3) *All constants and variables that are common to L_1 and to L_2 are over the same sorts in both signatures.*

In case L_1 and L_2 agree on their common symbols $L_1 \cap L_2$ and $L_1 \cup L_2$ can be formed. The former is the common signature and the latter the unified signature.

Now suppose that L_1 and L_2 are signatures that agree on their common symbols, and let $L = L_1 \cup L_2$ be their unified signature. Let S_1 and S_2 be two systems in signatures L_1 and L_2 respectively. (As defined above, the term *system* refers to a collection of system executions in some signature.) Then we define

$$S_1 \| S_2$$

as the system consisting of all system executions S in signature L such that $S|L_1 \in S_1$ and $S|L_2 \in S_2$. That is, a system execution is in $S_1 \| S_2$ iff its reduct to L_i is in S_i for both $i = 1, 2$.

From a slightly different angle $\|$ can be interpreted as intersection. If T_1 is a system for some signature G_1, and G_1 is a sub-signature of G_2, then one can naturally extend T_1 to a G_2-system obtained by expanding all system executions in T_1 in all possible manners to system executions for G_2. That is, one can define for any system execution S in G_2, $S \in T_1^{G_2}$ iff $S|G_1 \in T_1$. With this notation we get

$$S_1 \| S_1 = S_1^{L_1 \cup L_2} \cap S_2^{L_1 \cup L_2}.$$

3. Higher and lower-level events

An advantage of using system executions is the facility of working at different levels of granularities involving both higher-level and lower-level events. A higher-level event may be revealed to consist of minute lower-level events at a finer analysis. We begin with an example.

Suppose a *Writer* and several *Readers*, where *Writer* can write onto register R and any *Reader* can read it. Register R comprises in fact two relay stations, Relay1 and Relay2, which are used by the write and read operations. The write-operation "write(R, v)" (write variable v onto register R) is performed by making successive (telephonic) connections from the location of *Writer* to Relay1 and to Relay2 (in any order) and depositing the value v in these stations. The reading is done by connecting a single relay station (the first to which there is a connection) and obtaining the value. Such an arrangement may have the advantage of quicker access for the *Reader* in case of busy lines. Now look at the following scenario: Register R carries the value 0 in its two relays and *Writer* changes it to 1 by first calling Relay1 and then Relay2. In between these two calls, *Readers* are doing two read operations, r_1 and r_2, one after the other. The first read, r_1, calls Relay1 and obtains the new value 1. The second read, r_2, finds the line to Relay1 busy and obtains Relay2 before it was updated by the *Writer* and returns the value 0. That is, even though 1 was written after 0, to the *Readers* it seems as though 0 was in the register after 1. This somewhat paradoxical situation is called the *inversion phenomenon*, and it is typical of a certain type of register called by Lamport "regular".

Suppose now that *two Writers* are allowed to access Relay1 and Relay2. It is possible then for the following anomaly to happen: two write operations by the two *Writers* are interleaved in such a way that Relay1 obtains the value 0 and Relay2 the value 1. Thus different reads may get different values depending on which relay they access. (Some authors think that such a device is not a register at all, but we are interested here in modeling communication rather than finding the right definition of the term register.)

What signature should be chosen to model this situation?

- The lower-level events will be the read/write connections to the relays. So we should have predicates *Read* and *Write*, where $Read(x)$ and $Write(x)$ indicate that x is a lower-level read or a write event respectively. Then we want predicates *relay1* and *relay2* which when applied to a lower-level communication event tell us with which relay the communication is done, and finally we need a function *Value* to give the values written or returned by the operations. These values must then be of a specific sort.

- The higher-level events are the *Writer*'s operations which comprise of two accesses to the relays. So each higher-level event W is a set $W = \{w_1, w_2\}$ of lower-level write events to the two relays.

- We also need to know when a lower-level event a is part of a higher-level event A. Viewing a higher-level event as a set of lower level events, we use the set theoretical notation $a \in A$.

- The temporal precedence relation $\prec$ on the higher-level events is determined by its restriction to the lower-level events. It is clear for example that $\{w_1, w_2\} \prec x$ holds iff both $w_1 \prec x$ and $w_2 \prec x$ hold.

Having this example in mind, let us return to the general discussion. A priori there is no reason to stop after two levels, and one may think about the possibility of grouping higher-level events to form still higher-levels. In this book however, we shall never have the need for more than two levels at once. The situation where both higher-level and lower-level events appear is so common that the predicates *Higher_Level* and *Lower_Level* are introduced as standard predicates in every system execution. In addition, the membership relation $a \in A$ is used to relate a lower-level event a and a higher-level event A in which it occurs. When this is done, the following properties should be added to the definition of system executions.

(1) For every events a and A, $a \in A \rightarrow Lower\_Level(a) \wedge Higher\_Level(A)$.

(2) If X is a higher-level event that contains a nonterminating event then X is nonterminating as well. That is $\exists x (x \in X \wedge \neg T(x)) \rightarrow \neg T(X)$.

   If X is a higher-level event and X only contains terminating lower-level events, then X is terminating if and only if X is finite.

(3) For all higher-level events A and B

$$A \prec B \text{ iff } \forall a \in A \forall b \in B \ (a \prec b).$$

Similarly, if A is a higher-level event and b a lower-level event then

$$A \prec b \text{ iff } \forall a \in A \ (a \prec b), \text{ and}$$

$$b \prec A \text{ iff } \forall a \in A \ (b \prec a).$$

In a typical situation we are given a system execution with a set of lower-level events E, and we define higher-level events by some formula with parameters. For example, if the formula is $\phi(x, a)$ where x is a free variable and a a parameter, then for any value of a the collection of all x's that satisfy $\phi(x, a)$ is a higher-level event.

One must be careful to guarantee that the finiteness property holds after the addition of these higher-level events. For example, suppose that $a_1 \prec a_2 \cdots$ is an infinite sequence of lower-level events, and $E_i = \{a_1, a_i\}$ are formed. Then $\{E_i | i \in \mathbb{N}\}$ is an infinite antichain of terminating events! If on the other hand the higher-level events introduced are pairwise disjoint, then the finiteness condition automatically holds, as can easily be seen. In fact, a necessary and sufficient condition for the finiteness property is that each lower-level event belongs to only a finite number of higher-level events.

3.1. Beginning of events. It is often convenient to be able to refer to the start (and end) of an event. If A and B are higher-level events and for some lower-level events $a \in A$ and $b \in B$ $a \prec b$, then $begin(A) \prec begin(A)$ naturally holds. One possibility to interpret this relation is by the introduction of special "begin" events to which $begin(X)$ refers. The interpretation adopted here is to view

$$begin(X) \prec begin(Y)$$

as an abstract relation (which can be added to any system execution) that satisfies the following:

(1) Transitivity and irreflexivity hold: If $begin(X) \prec begin(Y)$ and $begin(Y) \prec begin(Z)$ then $begin(X) \prec begin(Z)$, and it is never the case that $begin(X) \prec begin(X)$.

(2) If for some event e $X \vdash e \prec Y$, then $begin(X) \prec begin(Y)$. (Hence, in particular, $X \prec Y$ implies $begin(X) \prec begin(Y)$.)

If $X \prec Y$ and $begin(Y) \prec begin(Z)$ then $X \prec Z$ follows. Indeed, assuming the premises, but supposing $Z \vdash X$, $begin(Z) \prec begin(Y)$ follows. So $begin(Z) \prec begin(Z)$ by transitivity, which is impossible. Assuming these properties of $begin$, one can show that representations exist which respect the $begin(X) \prec begin(Y)$ relationship.

4. Specification of registers and communication devices

System executions are used to formally specify communication devices. In this section we show first how to specify registers and then general communication devices.

A register is a communication device that accommodates one or more writing processes and one or more reading processes. A register is like a shared memory cell which can be changed by the writer(s) and which the reader(s) can access without modifying it. Thus a queue is not a register because a GET operation on a queue changes its contents. A register is associated with a range of values which it can hold. Informally, we say that a register "acquires" a value as a result of a write operation of that value, and we think of this write as explaining the results of subsequent reads. I prefer however to use a more abstract approach, in which registers are not objects to which values are assigned. Values are rather assigned to read/write events, and the specification of the registers determines the relationship between these values. We shall show how to specify registers and give several examples: serial, safe, and regular registers will be defined first. Then, to exemplify the possible diversity, throw/catch registers will be defined.

At this stage, programs and procedures are understood intuitively. A process is an execution of a program (or a protocol), and several processes can run concurrently. Among the program's instructions we may find read and write operations on registers, and connected with each register is its type of values that can be used for reading and writing. We distinguish between variables and registers. If v is a variable and τ an expression, then "$v := \tau$" is the instruction to assign value τ to v, but if V is a name of a register then some special instruction such as $Write_V(\tau)$ (rather than $V := \tau$) is used to write τ onto V. Similarly, a special instruction such as $Read_V(x)$ is used to assign to a (local) variable x a value obtained by reading register V. A formal definition of protocol languages and a more careful distinction between external operations (such as read/write on registers) and assignment instructions is given in Chapter 3. Here we are interested in the semantics of registers independently of their usage in programming. We ask: what does it mean to say that a register (or any communication device) is operating correctly?

To define the semantics of a register R we fix a signature L consisting of:

(1) Two sorts: "events" and "values".
(2) Two unary predicates: $ReadEvent_R$ and $WriteEvent_R$ are defined on the events. The idea is that if $ReadEvent_R(r)$ holds in some system execution then r is a read-of-R event, and if $WriteEvent_R(w)$ holds then w is a write-on-R event.

   We use here two sets of symbols: $Read_R/Write_R$ are instructions (in some computer language) and $ReadEvent_R/WriteEvent_R$ are predicates. Some other notations could also be used of course. In a casual discussion, one may prefer to use fewer symbols and to use $Write_R$, for example, both as an instruction and as a predicate. Then one relies on the context to make the instruction/predicate distinction.

(3) A function $Value$ is defined on the $Read_R/Write_R$ events, with range in the values sort. If r is a read-of-R event, then $Value(r)$ is called the value "returned" by r, and if w is a write-on-R event then $Value(w)$ is the value "written" by w.

(4) A unary function symbol ω_R is defined on the read-of-R events, and its values are write-on-R events. If $w = \omega(r)$ then we say that read r "obtained" the value written by w. (It is for graphical clarity that we often omit the subscript R from ω_R.) We say that ω is the "return" function

(5) Unary predicates $P_1, \ldots P_m$ are defined on the events. These predicates are called "processes", and if $P_i(e)$ holds then we say that event e is "executed" by process P_i.

In the first-order language derived from the signature L one can write properties of read/write events and their values which express the desirable, correct operations of a register. If τ (a sentence of L) is such a property, then the collection of system executions that satisfy τ describes, by definition, those system executions where the register is operating properly.

DEFINITION 4.1. *A specification for a register R is a system S for the signature L described above. We say that a system execution $\mathcal{M}$ describes a correct operation of register R if and only if $\mathcal{M} \in S$.*

Several examples will elucidate this (meta)definition.

DEFINITION 4.2. Serial registers *are characterized by the following axioms.*

(1) *The set of read/write events on R are linearly ordered by $\prec$, and the first event is a write event (called the initialization write-on-R).*
(2) *For all events r, if $Read_R(r)$, then*
    (a) *$Write_R(\omega(r))$,*
    (b) *$Value(\omega(r)) = Value(r)$,*
    (c) *$\omega(r) \prec r$ and there is no w such that $\omega(r) \prec w \prec r$ and $Write_R(w)$.*

The system S that specifies serial registers is then defined to be the collection of all system executions that are models of these two axioms.

Serial registers are usually defined by requiring that every read returns the value of the last write that preceded it. The existence of the initial write implies that any read r has a write w such that $w \prec r$, and the finiteness condition ensures that a last (maximal) such w can be found (the seriality implies uniqueness of this last w). This is a legitimate and equivalent approach, but I find that the use of a return function ω provides a greater flexibility of expression.

What is the role of the process predicate P_i ? Registers can be classified according to which processes can access their read/write operations:

(1) Single writer, single reader (or one-to-one) registers connect two processes. Some P_i is the *Writer* process and another is the *Reader* process (so if w is a write event then it is in *Writer*, and if r is a read event then it is in *Reader*). In this case $P_1 = Writer$, $P_2 = Reader$ are predicates with an obvious meaning. *Writer(e)* for example means that event e "belongs" to writer.
(2) Single writer, multiple reader (or 1-to-k) registers have a single writer process and several readers.
(3) Multiple writer, multiple reader registers can be accessed for writing by any process from the *Writer* list, and can be accessed for reading by any process from the *Reader* list. (It is not impossible for a single process to be in both lists.)

Observe that in this explication of the term register there is no object that can be called a register. No individual, predicate or function of a structure is a register, and there is no need to assign values to registers. Only read and write events have values and the specification of registers is abstractly determined with systems of structures.

DEFINITION 4.3 (SAFE AND REGULAR REGISTERS). Safe registers *are registers for which it is possible for write and read events to overlap. In such a case the value returned by the read is determined arbitrarily and it is not correlated*

with the value of an overlapping write or any other write. The only require-
ment is that the writes are linearly ordered and if a read does not overlap any
write then it returns the value of the last write preceding it. Formally the safety
requirements for a register R are:

(1) *The write events on R are linearly ordered, with an initial write that*
    precedes any other event.
(2) *The function ω is defined on the read events and it returns write events.*
    For any read-of-R event r, $\omega(r)$ is the $\prec$-rightmost write-on-R event w
    such that $w \prec r$. (There is always such a write.)
(3) *One of the following two possibilities holds:*
    (a) *r overlaps with some write event. That is, there is a write-on-R event*
        w such that $\neg(w \prec r \lor r \prec w)$.
    (b) *$Value(\omega(r)) = Value(r)$.*

Regular registers *resemble safe registers in that reads and writes may overlap,*
but in case of an overlap, a read must return the value of some write with which
it overlaps (in distinction to a safe register which may in such a case return any
value of the appropriate sort). The formal specification of regular registers is
given as a conjunction of axioms.

(1) *The write events on R are linearly ordered with an initial write that pre-*
    cedes any other event.
(2) *The function ω is defined on the read events and it returns write events.*
    For any read r of R the following holds:
    (a) *$\neg(r \prec \omega(r))$, and there is no write event w on R such that $\omega(r) \prec$*
        $w \prec r$.
    (b) *$Value(\omega(r)) = Value(r)$.*

The system S of all system executions that are models of these two axioms defines
the notion of regular register. That is, given any system execution S in this
signature (or a larger signature) we say that the read/write events describe a
regular register iff S satisfies the two axioms.

Safe and regular registers were defined by Lamport [20] (my definition, I
believe, is essentially the same).

4.1. Specifying communication devices. Communication devices with
their external operations are used by the processes to obtain information from
the outside and to change the environment. In particular they are used for
interprocess communication. We say "external operations" to distinguish them
from "internal operations" (such as addition of two numbers) which are not
concerned with the environment. When we just say "operation" we usually
mean external. Here are some examples of devices with their operations:

- If device R is a register, then $Read_R$, $Write_R$ are its operations as specified above. $Read_R(x)$ obtains the value of R and assigns it to variable
  x, and $Write_R(\tau)$ affects the environment and sets the value of R to the
  value of expression τ.
- If M is a light-meter, then $Read_M$ is an operation designed to report the
  present reading value. It is used in the form $Read_M(f, t, x)$ where f is a

color constant (such as "red") saying which filter to use, and t is a term that gives the aperture duration (in seconds), x is a variable on which the average intensity value is returned.

- If device S is a semaphore, then $P(S)$ and $V(S)$ are the operations on S. ($P(S)$ and $V(S)$ are seen as one symbol each, representing operations with no parameters).
- If device Q is a queue, then PUT_Q and GET_Q are its operations. These are single-place operations: $PUT_Q(v)$ requires a parameter v for the value to put, and $GET_Q(x)$ returns the value on variable x (if the operation is successful).

Now we shall define what it means to specify a communication device in general. A communication device D comprises of a finite set $R_1, \ldots, R_k$ of external operations. The specification of D and its operations consists of two parts. The first part relates to the syntax of the instructions, and the second part to the events that are generated by executions of these instructions. It belongs to the syntactical part of the specification to say, for example, that $Read_R(5)$ is an illegitimate call, because $Read_R$ expects a variable rather than a constant. It belongs to the semantical part to require, for example, that any read returns the value of the last preceding write.

Let D be any device consisting of external operations $R_1, \ldots, R_k$. The first part (the syntactical part) of the specification requires the following:

(1) A structure B for some first-order language L_0. We say that B is the *background structure* for device D. The intention is that the universe of B contains the data types (sorts) of the parameters of the external operations of D.

(2) For every operation R_j in D, a natural number $r_j \geq 0$, called the arity of R_j, and a vector v of length r_j of pairs of the form $\langle \text{variable}, X \rangle$ or $\langle \text{value}, X \rangle$ where X is a sort in B. If $v[i]$ (the ith value of v) is $\langle \text{variable}, X \rangle$, then we say that the ith entry (or coordinate) of R_j is a variable parameter of sort X, and if $v[i] = \langle \text{value}, X \rangle$ then we say that the ith entry is a value parameter of sort X.

When R is one of the operations R_j, we use the notation $R(t^1, \ldots, t^k)$ to denote that the arity of operation R is k and that the mth entry of R is of kind t^m where t^m has the form $\langle \text{variable}, X \rangle$ or $\langle \text{value}, X \rangle$ with X a sort in B. The kind t^m thus includes the information of whether m is a variable or value place, and what sort of values is expected at this place. For example,

$$Read_R(\langle \text{variable}, \text{integer} \rangle)$$

indicates that operation $Read_R$ is unary and it accepts an integer variable. The structure B in this case would be the familiar structure of the integers. Another example involves an operation with three parameters:

$$Read_M(\langle \text{value}, \text{color} \rangle, \langle \text{value}, \text{duration} \rangle, \langle \text{variable}, \text{intensity} \rangle).$$

This says that $Read_M$ is a ternary operation, and that the three parameters that need to be supplied with each invocation of $Read_M$ are of type *color, duration, intensity*. These are sorts in some structure B_M; the first two parameters are value parameters while the third is a variable. We have in mind a light-meter that gets an instruction to read the intensity at a certain wave length for a certain period, and reports the result on some variable.

There exist natural examples in which a device consists of more than two operations. For example a test-and-set register may support a write operation (with a single value-parameter), a read operation (with a single variable), and a conditional write (with two value-parameters a and b with the intention that only if the value of the register is a does the operation change it to b).

Now the second part of the specification of a device defines its semantics. It is given with system executions that describe the correct behavior of the device. Let B be the background structure of D in language L_0, and for each R_j let r_j be its arity and $R_j(t_j^1, \dots, t_j^k)$ specify the kind of each entry of R_j. Usually, the specification of D is given by means of some sentences that describe the desirable properties of the operations. More precisely, the specifying sentences describe the events generated by the operation executions. Instead of sentences, we speak in the following definition about the models of these sentences, namely about collections of system executions.

DEFINITION 4.4 (DEVICE SPECIFICATION). *A specification for a communication device D consists of a first-order signature L_D and a system S_D (a family of system executions in L_D) such that the following hold.*

(1) *L_0 (the language of the background structure) is a sub-signature of L_D. For each operation symbol $R = R_j$, L_D contains a unary predicate R-event defined on the events. If R-event(e) then e is said to be an execution of operation R.*

(2) *L_D contains a function symbol Value. Value is defined on the events and it gives a sequence of values for every event that is an execution of some R_j operation. Let $R = R_j$ be any operation of arity r and kind given by $R(t^1, \dots, t_k)$. Let R-event be the corresponding predicate. The following then holds in any system execution S in S_D. For every event e in S, if R-event(e) then $Value(e) = \langle a_1, \dots, a_k \rangle$ is a k-tuple of values such that a_i is in B of the sort given by t^i.*

(3) *The background structure B is a reduct of every system-execution in S_D.*

(4) *L_D may contain more predicates and functions which may be used in the specification of D.*

If R is an operation such that every execution of R in any system execution in S_D is terminating, then we say that R is a terminating operation. The serial, safe, and regular register specifications given above are examples of device specifications. We shall give now another example: A regular register D that supports read/write operations on real numbers, but in such a way that the value returned by a read is a rational number whose distance from the written real value does not exceed 0.001. We first define L_D.

(1) The background structure B and its language L_0 contains two sorts—the real numbers and the rational numbers—and the usual relations, functions, and constants that are associated with the reals. (That is, $<$, $\times$, absolute value $|\cdot|$ etc.)

(2)

$$Read(\langle\text{variable}, \mathbb{Q}\rangle), Write(\langle\text{value}, \Re\rangle).$$

This is the syntactical information saying that D consists of two unary operations, *Read* and *Write*, where $Read(x)$ can be invoked with a variable x that acquires a value which is a rational number, and where *Write* accepts a single value parameter of type real. (So $Write(\pi)$, for example, is a legitimate call, if π is a constant in the language.)

(3) The specification language L_D contains, in addition to L_0, two predicates, *Read-event*, and *Write-event*, defined on sort *Event*. L_D also contains a function, *Value*, defined on the events. If *Write-event(e)*, then $Value(e)$ is a real number (the value "written") and if *Read-event(e)* then $Value(e)$ is a rational number (the value "returned").

(4) L_D contains a function ω (called the return function) defined on the events.

As in Definition 4.4, the specification of D is a system S_D in L_D defined as follows. L_D consists of all system executions in signature L_D that satisfy the following axioms.

(1) The write events are linearly ordered with an initial write that precedes any other write:

$$\forall e_1, e_2 \; [\textit{Read-event}(e_1) \wedge \textit{Read-event}(e_2) \wedge e_1 \neq e_2 \rightarrow$$
$$e_1 \prec e_2 \vee e_2 \prec e_1]$$

$$\exists e(\textit{Write-event}(e) \wedge \forall r(\textit{Read-event}(r) \rightarrow e \prec r))$$

(2) For any read event r, $\omega(r)$ is the $\prec$-rightmost write event preceding r.

$$\forall r[\textit{Read-event}(r) \quad \rightarrow \quad \textit{Write-event}(\omega(r)) \wedge \omega(r) \prec r \wedge$$
$$\neg\exists e(\textit{Write-event}(e) \wedge \omega(r) \prec e \prec r)]$$

(3) For every read event r, either

$$\exists e(\textit{Write-event}(e) \wedge \neg(e \prec r) \wedge \neg(r \prec e))$$

or

$$|Value(r) - Value(\omega(r))| < 0.001$$

Example: *throw/catch.* The following device, consisting of *throw/catch* operations will be used later on; it is brought here as an additional detailed example.

Imagine two players, *Pitcher* and *Catcher* playing the following game. *Pitcher* throws balls containing messages in the air, and *Catcher* tries to catch and put them aside whenever the catch is successful. Since *Pitcher* throws the balls to different heights, it is possible for the catcher to receive balls not in the order they were thrown and even to miss some of them. A catch operation may be nonterminating (either because no more balls are thrown, or because the catcher is looking in the wrong direction). A ball may duplicate and multiple balls carrying the same original message may result from a single throw.

To model this situation we define a signature L consisting of the following:

(1) Two unary predicates: *catch-event* and *throw-event*. If *catch-event*(c) holds then c is called a catch event, and if *throw-event*(t) then t is a throw event. *Terminating* is a standard predicate, defined on the events, and $\prec$ is the standard precedence relation defined on the events.

(2) A sort "Data-values" is specified in L, and a function $Value : Events \rightarrow$ Data-values is defined on all the terminating throw/Catch events.

(3) A unary function symbol α denotes a function defined on the terminating catch events. $\alpha(c)$ is a throw event. If $t = \alpha(c)$ then we say that catch c "obtained" the value thrown by t.

The single axiom that determines the *Throw/Catch* system is the following:

If *catch-event*(c) and *Terminating*(c) then $\alpha(c) \prec c$
and $Value(\alpha(c)) = Value(c)$.

(Since $\alpha(c) \prec c$, $\alpha(c)$ is necessarily a terminating throw event.) Let $S_{throw/catch}$ denotes the collection of all system executions (of signature L) that satisfy this axiom.

We shall return to the throw and catch operations in the following chapter.

As an exercise, the reader can try to write a specification for a light meter that measures the red light intensity with accuracy of 2% and the blue light with only 4%, if the aperture is for 1 second and the intensity is within a certain range $0 < a < b$ (see also the light-meter example on page 8). The background structure for such a specification has to enable arithmetical expressions.

5. Achilles and the Tortoise

Any theory of concurrency has to explicate (to formalize) the intuitive notion of concurrent behavior. It may be illuminating to see how such a theory deals with situations that are not from the realm of computer science, to check how flexible and adaptive the theory is. I hope that the reader will enjoy the following exercise in applying system executions to the Achilles paradox of Zeno.

Usually events have positive (i.e., > 0) duration. Yet even in everyday experience we often associate a single number to record the time of an event, rather than a pair of numbers to indicate an interval. For example, we say "Achilles finished the race at $14 : 34 : 21$" to record his arrival time.

Henceforth we do accept durationless events (in addition to usual ones), and call them "instantaneous events". The time of an instantaneous event e is denoted *time*(e). Usually real numbers represent time instants, but other choices are possible for a temporal domain. We require here that the temporal domain is linearly ordered, and we write $t_1 < t_2$ for temporal instants t_1, t_2 when t_1 comes before t_2.

$\prec$ denotes the precedence relation on the events. It is not a linear ordering since two events may overlap, but the restriction of $\prec$ to the instantaneous events *is* linear and

$$e_1 \prec e_2 \text{ iff } time(e_1) < time(e_2).$$

Recall the finite precedence property which says that the set of events that precede any given event is finite. That is, for any event e the set

$$\{x \mid x \text{ is an event and } x \prec e\}$$

is finite. Although the set of instants may be infinite and dense, we accept this finiteness demand for the instantaneous events (and for all other events).

Only terminating events are considered in this section, that is events with a finite duration.

It follows immediately from our discussion that the set of instants

$$\{time(e) \mid e \text{ is an instantaneous event}\}$$

is at most of order-type ω (the order type of the natural numbers). So, if the domain of instants is dense, then surely there are instants that are not of the form $time(e)$. These are mathematical objects that do not correspond to any "real" event.

It makes sense however to assume that every *definable* instant is of the form $time(e)$ for some instantaneous event e. That is, definable instants have a higher existential status under this assumption. Accepting this view puts a severe limitation on the language in which these instants are defined, for if the language is too rich then there would be definable instants with no corresponding instantaneous events. For example, if division by 2 is allowed, then the so called "dichotomy" argument shows that there must be definable instants with no corresponding events. Indeed, assuming the existence of two instantaneous events with times $t_0 < t_1$, define first $t_2 = (t_0 + t_1)/2$ and continue to define the midpoints until you reach an instant with no corresponding event (for there are only finitely many events preceding the event of t_1). It can be argued however that "division by 2" is not a natural operation in this context, since the fact that both t_0 and t_1 are remarkable moments does not say anything about their midpoint. The Achilles paradox is presented here as an argument that, even in a very simple language with no arithmetical operations, concurrency brings about a definable event-less instant.

We are used today to represent both time and space with real numbers, and by this we implicitly make quite a strong statement about the nature of time and space, namely their isomorphism. Since the argument of Zeno in the Achilles paradox does not assume this time–space correspondence, we will not impose it and use two sorts R_χ and R_τ to represent time and location (chronos and topos).

The following defines a signature, called L_A (A is for Achilles), which we use to describe Achilles' runs.

(1) Three sorts: E (for events), R_χ (for instants), and R_τ (for locations).

(2) A binary relation symbol $\prec$. Its domain is specified to be E. $\prec$ is called the "precedence" relation on the events. (When we say that the domain of $\prec$ is E, we mean that the semantics of the formula $x \prec y$ imply that both of x and y are in E.)

(3) Two binary relation symbols $<_\chi$, and $<_\tau$. Their domains are R_χ and R_τ respectively. However, for simplicity of expression, we will write $<$ instead of $<_\chi$ or $<_\tau$ when no confusion is possible. (The intention is to use $s < t$ when instant s is earlier than t, and to use $u < v$ for locations u and v when u is nearer than v to the starting point of the race.)

(4) A unary predicate symbol I, defined over sort E. (Our intention is to use $I(x)$ for "x is an instantaneous event".)

(5) A unary function symbol *time*. Its domain is sort E and its range is sort R_χ. (The value $time(e)$ is called "the instant of e". Of course, it is only meaningful if e is instantaneous, and yet *time* is required to be defined on all events, even if not instantaneous. This problem comes from the requirement that functions are total on their sort, and often a special "undefined" value is used to enable total functions to mimic partial ones.)

(6) A unary function symbol d_A has domain R_χ and range R_τ. It is called "the location function" and our intention is that $d_A(t)$ gives the location of Achilles at instant t.

(7) Two constants t_0^A and t_1^A are specified to be in R_χ. (t_0^A will be used to denote the time Achilles starts his course, and t_1^A is the time he finishes his course.)

Let $\mathcal{M}$ be an interpretation for L_A. So $\mathcal{M}$ assigns to each symbol X in L_A an object $X^{\mathcal{M}}$ such that:

(1) $E^{\mathcal{M}}$, $R_\chi^{\mathcal{M}}$, and $R_\tau^{\mathcal{M}}$ are non-empty pairwise disjoint sets. (Members of $E^{\mathcal{M}}$ are called events of $\mathcal{M}$, members of $R_\chi^{\mathcal{M}}$ are called instants, and members of $R_\tau^{\mathcal{M}}$ are called locations.)

(2) $\prec^{\mathcal{M}}$ is a binary relation on $E^{\mathcal{M}}$.

(3) $(<_\chi)^{\mathcal{M}}$ is a binary relation on $R_\chi^{\mathcal{M}}$, and $(<_\tau)^{\mathcal{M}}$ is a binary relation on $(R_\tau)^{\mathcal{M}}$.

(4) $I^{\mathcal{M}}$ is a unary relation on $E^{\mathcal{M}}$, that is a subset of $E^{\mathcal{M}}$.

(5) $time^{\mathcal{M}} : E^{\mathcal{M}} \longrightarrow R_\chi^{\mathcal{M}}$. That is, $time^{\mathcal{M}}$ is a (partial) function assigning instants to events.

(6) $(d_A)^{\mathcal{M}} : R_\chi^{\mathcal{M}} \longrightarrow R_\tau^{\mathcal{M}}$.

(7) $(t_0^A)^{\mathcal{M}}$, $(t_1^A)^{\mathcal{M}} \in (R_\chi)^{\mathcal{M}}$.

For simplicity of expression, we may omit $\mathcal{M}$ and write, for example, E instead of $E^{\mathcal{M}}$.

Since we are interested in structures that represent possible runs of Achilles, certain interpretations may be discarded at once. For example, those in which $<$ is not a linear ordering on R_χ. The collection of all "acceptable" structures for L_A will be denoted S_A, but how are we going to define it? There will be two tests an L_A-structure must pass in order to be included in S_A: first it must be a system execution, and second it has to be a model of the following theory Th_A. That is, S_A is defined to be the class of all system executions that are models of Th_A; it is called the *system* for Th_A.

Th_A is the following finite set of first-order sentences in the language of L_A (for simplicity some of these sentences are given in semi-formal English).

(1) $<_\chi$ and $<_\tau$ are irreflexive linear orderings on R_χ and R_τ respectively. (We will use the plain symbol $<$ for both $<_\chi$ and $<_\tau$, and, as usual, $x \leq y$ is a shorthand for $x < y \vee x = y$.)

(2) $t_0^A < t_1^A$, and the function d_A is monotonically increasing on the interval $[t_0^A, t_1^A]$. That is

$$\forall r, s \in R_\chi \; (t_0^A \leq r < s \leq t_1^A \longrightarrow d_A(r) < d_A(s)).$$

(3) $\forall e_1, e_2 \in E \; [I(e_1) \wedge I(e_2) \longrightarrow ((e_1 \prec e_2) \text{ iff } time(e_1) < time(e_2))]$.

(That is, for any two instantaneous events the precedence relation $\prec$ and the $<$-ordering of their timings agree.)

The collection of all possible runs of the tortoise can be similarly defined. First we set the language by defining a signature L_T obtained by exchanging "A" with "T" in the symbols of L_A. That is, L_T contains:

(1) Three sorts: E, R_χ, and R_τ.
(2) A binary relation symbol $\prec$. Its domain is specified to be E.
(3) Two binary relation symbols $<_\tau$ and $<_\chi$ with domains R_τ and R_χ.
(4) A unary predicate symbol I, defined over sort E.
(5) A unary function symbol $time$. Its domain is sort E and its range is sort R_χ.
(6) A unary function symbol d_T has domain R_χ and range R_τ. (It is called "the location function for the tortoise" and our intention is that $d_T(t)$ gives the location of the tortoise at time t.)
(7) Two constants t_0^T and t_1^T are specified to be in R_χ. (t_0^T is used to denote the time the tortoise starts her course, and t_1^T is the time she arrives to the end.)

Now, Th_T is defined just as Th_A with the tortoise symbols instead of Achilles'. Then S_T is the collection of all system executions that satisfy the axioms of Th_T.

Intuitively then S_A and S_T represent the systems of all possible runs of Achilles and of the tortoise (separately).

5.1. Concurrency. Now we treat concurrency. What does it mean that both Achilles and the tortoise run concurrently? Disregarding possible psychological interdependencies, we represent such a concurrent run as an independent combination of a structure from S_A with a structure from S_T. Let us be more precise and define first a signature $L = L_A \cup L_T$. Specifically, L contains each of the symbols that appear in both L_A and L_T (such as $time$) with their specified arity, and then L contains d_A as well as d_T, and the four constants t_0^A, t_1^A, t_0^T, t_1^T. Now from any structure $\mathcal{M}$ for L, two structures can be derived: $\mathcal{M}|L_A$, an interpretation of L_A, and $\mathcal{M}|L_T$, an interpretation of L_T. (These are reducts of $\mathcal{M}$.) Then the concurrent system S is defined as the class of those structures $\mathcal{M}$ for which $\mathcal{M}|L_A \in S_A$ and $\mathcal{M}|L_T \in S_T$.

A general definition of reducts was given in Chapter 1, but we recall it for convenience. Let $\mathcal{M}$ be any structure for L, then the reduct $\mathcal{N} = \mathcal{M}|L_A$ is defined by taking the universe of $\mathcal{M}$ with its three sorts: $E^\mathcal{N} = E^\mathcal{M}$, $(R_\chi)^\mathcal{N} = (R_\chi)^\mathcal{M}$, and $(R_\tau)^\mathcal{N} = (R_\tau)^\mathcal{M}$. Then $\mathcal{N}$ interprets all the symbols in L_A exactly as they are interpreted by $\mathcal{M}$. In short, $\mathcal{N}$ is obtained by throwing away from $\mathcal{M}$ all interpretations of symbols not in L_A.

Similarly, a reduct to L_T restricts interpretations of L to interpretations of L_T. Note that the events in $\mathcal{M}$ pertaining to Achilles remain in $\mathcal{M}|L_T$ (since both structures have the same universe), but now they have "lost their meaning".

We now define

$$S_A \parallel S_T = \{\mathcal{M} \mid \mathcal{M} \text{ is a structure for } L \text{ such that } \mathcal{M}|L_A \in S_A \text{ and } \mathcal{M}|L_T \in S_T\}.$$

$S = S_A \parallel S_T$ is the system of all "concurrent executions" of Achilles and the tortoise. However, if $\mathcal{M} \in S$, then it is not necessarily a race: the point is that in a race two additional requirements must be satisfied.

DEFINITION 5.1. *We say that $\mathcal{M} \in S$ is a* race *iff*

(1) $t_0^A = t_0^T$ *holds in $\mathcal{M}$. That is, Achilles and the tortoise start the race at the same time, which we denote by t_0.*

(2)

$$(t_1^T < t_1^A) \vee \exists t \, [(t_0 < t \leq t_1^A) \wedge d_A(t) =$$
$$d_T(t) \wedge \forall s \, (t_0 \leq s < t \longrightarrow d_A(s) < d_T(s)].$$

That is, either the tortoise reaches the end of the course before Achilles does (and then she wins), or else there is an instant t, sometimes after the beginning but before (or at) the end of Achilles' race, in which Achilles reaches the tortoise for the first time: At each earlier instant the tortoise is ahead of Achilles (and in particular, then, at t_0, the tortoise is given a head start over Achilles).

Define R as the set of all "races", i.e., the set of all system executions in S that satisfy the two statements above.

We say that the tortoise wins in $\mathcal{M}$ if $t_1^T < t_1^A$ holds, but otherwise Achilles is declared the winner. That is, Achilles is the winner in $\mathcal{M}$ iff $t_1^A \leq t_1^T$ holds.

Definability has a central place in our interpretation of the Achilles paradox, and so we recall what it means for an individual (a member of the universe of a structure) to be definable. Let L be some signature and $\mathcal{M}$ a structure for L.

DEFINITION 5.2. *We say that a formula $\phi(x)$ (in the language of L and with a single free variable x) is a* definition *in $\mathcal{M}$ iff the following sentence holds:*

$$\exists ! x \phi(x).$$

The symbol $\exists ! x$ means as usual "there is a single x such that..." So, if $\phi(x)$ is a definition in $\mathcal{M}$ then there is a single individual a such that $\mathcal{M} \models \phi[x/a]$. This unique a in $\mathcal{M}$ is said to be "definable" by $\phi(x)$, and we say that a is $\mathcal{M}$-definable.

We now return to the signature $L = L_A \cup L_T$ defined previously and to the system S of all concurrent executions as defined above. We say that $\mathcal{M} \in S$ satisfies the *temporal existence property* iff for every definition $\phi(x)$ in $\mathcal{M}$ that defines an individual a in $\mathcal{M}$, if $R_\chi(a)$ holds in $\mathcal{M}$, then there is an event e in $\mathcal{M}$ such that $I(e) \wedge a = time(e)$.

Intuitively then, the temporal existence property says that any definable instant "really exists": a definable instant is necessarily experienced as the exact time of an event. This, I believe, is a reasonable property because it allows us to distinguish between those members of R_χ that are there just for theoretical completeness and those that have a "good reason" to exist.

Another definition is needed before we tackle the paradox itself. The image of a continuous function on a closed real interval is, as we all know, a closed interval contained between the maximum and minimum of the function. In particular, the intermediate value property holds for any continuous function: If $f(x_1) < y < f(x_2)$ then for some x between x_1 and x_2, $f(x) = y$. Functions in physics are usually assumed to be continuous, and hence they satisfy the intermediate value property. I do not know if the Greeks would accept continuity without hesitation, and if time is assumed to be discrete, then continuity is meaningless. Yet the intermediate value property of a physical function seems to be "evident". For example, if Achilles is at location d_1 at some moment and then at some farther location d_2 at a later moment, then surely he passes through every intermediate location. We base Zeno's paradox on a weaker property.

DEFINITION 5.3. *Let M be a system execution in S. We say that d_A satisfies the weak intermediate value property in M (in the temporal interval $[t_0^A, t_1^A]$) iff whenever a, in $(R_\tau)^M$, is M-definable and satisfies $d_A(t_0^A) \leq a \leq d_A(t_1^A)$ in M, then for some instant t in the interval $[t_0^A, t_1^A]$ in M, $d_A(t) = a$ holds.*

In other words, the weak intermediate value property says that even if the range of d_A omits some values, it obtains all those that are definable in M in the interval between $d_A(t_0^A)$ and $d_A(t_1^A)$ (the two extremes of this monotonic function in M).

5.2. Zeno's paradox. The usual description of the Achilles paradox is to assume constant speeds for Achilles and for the tortoise, to fix a head start that puts the tortoise in front of Achilles, and then to define an infinite, increasing sequence of instants $\langle t_i \mid i = 0, 1, \ldots \rangle$ such that for each i the tortoise at t_i is strictly between Achilles and the finish post. To begin with, t_0 is the starting time of the race. If t_i is defined and $l_i < l_i'$ are the locations of Achilles and the tortoise respectively, at t_i, then t_{i+1} is computed as the time by which Achilles reaches l_i'. The final argument derives a contradiction from the contrast between this non terminating description and the termination of the race.

I wish to find the simplest possible first-order language in which the paradox can be presented, and for this a somewhat different path of reasoning is chosen in which the speeds of Achilles and the Tortoise are irrelevant.

THEOREM 5.4 (ZENO). *Suppose that M is a race in which Achilles wins. If the location function d_A in M satisfies the weak intermediate value property, then the temporal existence property does not hold. That is, there is a definable instant with no corresponding event.*

Proof. Suppose that Achilles wins the race M. This means by definition that $t_1^A \leq t_1^T$ holds. Assume that the temporal existence property holds, and we shall derive a contradiction to the finite predecessor property which holds in every system execution.

Since $<$ is irreflexive and transitive $t_1^A \leq t_1^T$ implies that $t_1^T < t_1^A$ does not hold, and thus by (2) in Definition 5.1 there exists an instant t' such that the following holds in M

(6) $t_0 < t' \leq t_1^A \wedge d_A(t') = d_T(t') \wedge \forall s(t_0 \leq s < t' \longrightarrow d_A(s) < d_T(s)).$

Let

$$f = d_A(t') = d_T(t')$$

denotes the location in $\mathcal{M}$ where Achilles reaches the tortoise for the first time. Observe that t' is definable because there is at most a single individual t' in $\mathcal{M}$ as above. (For if t'' is another such instant then $t' < t''$ or $t'' < t'$ by the assumed linearity of $<$. If $t'' < t'$ for example, then $s = t''$ contradicts $\forall s(t_0 \leq s < t' \longrightarrow d_A(s) < d_T(s))$).

Hence by the assumed temporal existence property for some instantaneous event called "catch_up_event",

$$t' = time(catch\_up\_event).$$

We are going to define an infinite series of definable instants t^i, for $i \in \mathbb{N}$ (the set of natural numbers) such that $t_0 \leq t^i < t^{i+1} < t'$ holds for every i. But this is impossible since the temporal existence property guarantees an instantaneous event e_i such that $t^i = time(e_i)$ for each i, and hence that the e_i's are all distinct and precede $catch\_up\_event$.

$t^0 = t_0$ is defined as the start of the course, and $t_0 < t'$ follows from (6). Suppose that t^i has been defined with $t_0 \leq t^i < t'$. Then

$$d_A(t^i) < d_T(t^i) < f$$

by formula (6) and by the monotonicity of d_T. Since t^i is definable, $a = d_T(t^i)$ is definable as well, and by the above formula

$$d_A(t_0) \leq d_A(t^i) < a < d_A(t') = f.$$

so the weak intermediate value property can be applied to yield an instant t^{i+1} such that $t^i < t^{i+1} < t'$ and $d_A(t^{i+1}) = a$. The monotonicity of d_A implies that t^{i+1} is definable as the unique instant s in the interval $t_0 < t_1^A$ for which $d_A(s) = a$. $\square$

Though system executions were defined with the second-order finiteness property, the reader may have remarked that we only used a first-order property in the proof: that for every event, if it is preceded by an instantaneous event, then it is preceded by a last (i.e. $\prec$-maximal) instantaneous event. In general, the finiteness property can be replaced by a scheme of first-order axioms.

3
Semantics of concurrent protocols

The behavior of concurrent protocols is often difficult to follow, even when they are simple and short. In this chapter a simple protocol language is defined and its semantics is given by means of local states and system executions. The purpose of this semantics is to be sufficiently clear and manageable not only to support impeccable correctness proofs, but also to correlate with our intuitive understanding. A key idea is to split the semantics in two parts: internal and external.

There are two aspects to semantics, which I call internal (or unrestricted) and external (or restricted), that should be discussed before getting on to the technical issues. Usually these aspects are intertwined, but here they are separated and hence understanding the difference is a key issue.

By *internal semantics* I mean analyzing the *forms* of programs and relating these forms to meaning, without considering the external world and the communication devices. For example, it is part of internal semantics to define that the execution of $P_1; P_2$ is done by first executing P_1 and then (if it terminates) executing P_2 in such a way that the starting state is the state in which P_1 has terminated. The notion of state is of prime importance here, but only local states influence the internal semantics. The term "unrestricted semantics" is also used because the communication devices are not restricted to operate properly: a read for example may return any value (within its range) even if that value has never been written.

By *external semantics* I mean the specification of communication devices and the relationship of the programs with the external environment. The semantics of registers, queues, or channels, for example, do not depend on the programs that use them, and ought to be described separately from the programs' semantics. In describing the semantics of communication devices and how it affects the programs, system executions are of prime importance. The term "restricted semantics" is also used to convey the idea that the communication devices operate properly.

There seems to be a consensus among computer scientists to describe concurrent systems by means of *states* and *transitions*. According to this model of concurrency a parallel execution of sequential procedures $P_1 \ldots, P_n$ is obtained as an arbitrary interleaving of atomic steps of the processes. Each step "taken" by some P_i changes the global state in accordance with well defined rules, and the manifold of all possible sequences of states thus obtained is the semantics of the concurrent procedures. We have seen an example of this in Chapter 1 Section 3. It seems to me that this interleaving view is inadequate for very complex environments in which the programs are embedded. In antithesis to this state/transition approach we use no global states here; only local states to describe the internal semantics, and then system executions to describe the external semantics (the environment and its interrelation with the program).

1. A protocol language and its flowcharts

Our plan is to describe a family of protocol languages, and then to define their internal and external semantics. For the internal semantics we use state-and-transition structures and histories. Once this is done, executions are defined which enable the analysis of external semantics as well. First we justify the use of background structures in defining languages, and recall how external operations were defined.

Distinct protocol languages may be different because of different choices of primitive terms and instructions. Thus before specifying a protocol language we must know what the meaning of its symbols, terms and formulas is. For this we assume a fixed first-order language L and a structure $\mathcal{B}$ for L, called the *background language and structure* for the protocol's language. Now when we see an assignment instruction such as $x := y + 5$ we know that the term $y + 5$ has to be evaluated in $\mathcal{B}$. Similarly, when a conditional instruction such as if $x < y$ then $Send(y)$ is encountered, we know in which structure to look for the truth value of $x < y$.

Remark: Often the condition $x < y$ is called a boolean expression, but we call it a formula, following model theoretic terminology which makes a distinction between terms (expressions) and formulas. (Terms have values in the structure and formulas are evaluated to True or False given some assignment).

The protocol language borrows from the first-order language L its terms and formulas. In fact only computable terms and formulas of L are used in a computer language, but we disregard this aspect here.

Included in the first-order language L is also a list of sorts, variables, and constants connected with each sort (see Definition 2.1). The sorts, variables and constants of the first-order language L are also the sorts, variables and constants of the protocol language (sorts are called types in this context). There is no "variable declaration" part in our programs, since we rely on this fixed declaration in L which assigns to each variable its type. Thus, for example, if i is a variable of sort integer in L, then it can be used in any program as a variable of type integer with no prior declaration. (This assumption simplifies somewhat our definitions, but clearly real programmers want the flexibility of choosing their own variables.)

Now, besides the standard key words (such as if or **then**) and the identifiers (which serve as names of procedures), we need a list of "external operations" that serve for communication or coordination, and relate the program to the exterior. For example $Send(x)$, $Write_C(x)$, and $V(S)$ are external instructions. As explained in Chapter 2 section 4.1, every external operation symbol has an arity $r \geq 0$, and for each integer $1 \leq m \leq r$ two factors are determined:

(1) The sort of the mth entry (it is one of the L sorts interpreted in the background structure $\mathcal{B}$).
(2) The kind of the mth entry: whether it is a value or a variable parameter.

An external instruction is always of the form $R(t_1, \ldots, t_r)$, or just simply R (with no parameters), where R is the external operation symbol and $r \geq 0$ is its arity; each t_i is a term (possibly just a variable). Value places can accept any terms of the type associated with that place, but variable places can only accept variables (again of the right type).

A concurrent protocol is a program

$$\textbf{concurrently} \ \ \textbf{do} \ P_1, \ldots, P_n \ \textbf{od}$$

where each P_i is a serial procedure defined in detail in the following section. More complex forms of protocols in which the procedures P_i themselves can be concurrent programs, or where one procedure can call another, are certainly useful, but we shall deal in this chapter only with the simplest forms.

1.1. Serial and concurrent procedures. Suppose that a first-order language L and a background structure $\mathcal{B}$ for L have been fixed, and that a list of external operation symbols is given with the additional information, as required above, on the sort and kind of each entry. The protocol language can now be constructed: first simple instructions are defined, then compound instructions, and finally protocols, which are composed of concurrently executing serial procedures.

Simple instructions are either "assignment" or "external" instructions:

(1) An assignment instruction s has the form

$$x := \mu$$

where x is a variable and μ is a term in L that has the same type as the type of x.
(2) An external instruction has the form

$$R(t_0, \ldots t_{n-1})$$

Where R is an external operation symbol of arity n and each t_i is a variable or a term in L. The type of t_i must be the type associated with the ith place of R. Moreover, if this place is a value place, then t_i is a term, but if it is a variable place then t_i needs to be a variable.

Serial instructions (also called instructions) can be simple instructions, as defined above, concatenated instructions, conditional instructions, or while loop instructions, as defined below:

(1) If s_1 and s_2 are instructions, then

$$(s_1; s_2)$$

is an instruction called the concatenation of s_1 followed by s_2. Parentheses are usually omitted, and begin-end pairs are sometimes used to group blocks of instructions.

(2) If s is an instruction and τ is a condition (a formula in L) then

if τ then s

is an instruction called a conditional instruction.

(3) If s is an instruction and τ is a condition then

while τ do s

is an instruction. It is called a while loop.

To sum up the definition of (serial) instructions we write it in BNF (Backus–Naur Form).

⟨instruction⟩	$\longrightarrow$	*⟨simple⟩* | *⟨concatenated⟩* | *⟨conditional⟩* | *⟨while⟩*
⟨simple⟩	$\longrightarrow$	*⟨assignment⟩* | *⟨external⟩*
⟨assignment⟩	$\longrightarrow$	$x := \mu$, where x is a variable and μ a term of the same sort.
⟨external⟩	$\longrightarrow$	$R(t_0, \ldots, t_{n-1})$, where R is an external operation symbol and the terms t_i are as defined above.
⟨concatenated⟩	$\longrightarrow$	*⟨instruction⟩* ; *⟨instruction⟩*
⟨conditional⟩	$\longrightarrow$	**if** τ **then** *⟨instruction⟩*, where τ is a formula in L
⟨while⟩	$\longrightarrow$	**while** τ **do** *⟨instruction⟩*, where τ is a formula in L

For each instruction s, $\mathcal{V}(s)$ is the set of all variables that appear in s, that is, in the terms and formulas in s, and the variable to the left of the assignment operator.

A serial procedure is a (serial) instruction which has been given a name. The naming is only for our own convenience, since a procedure cannot be invoked from within another procedure or from itself. Formally, a serial procedure has the form

procedure P begin s end

where P is the name of the procedure, and s is a (serial) instruction (the *body* of P). If P is a serial procedure with body s, then $\mathcal{V}(P) = \mathcal{V}(s)$ is called the set of variables of P.

DEFINITION 1.1 (NON INTERFERENCE). *We say that serial procedures* $P_1, \ldots, P_n$ *do not interfere with each other iff for every* $i \neq j$ $\mathcal{V}(P_i) \cap \mathcal{V}(P_j) = \emptyset$. *In other words,* P_j *never mentions a variable that* P_i *does.* [1]

If $P_1, \ldots, P_n$ do not interfere with each other, then the *concurrent protocol*

concurrently do $P_1, P_2, \ldots, P_n$ od

can be formed. Such a protocol is said to be a concurrent combination of serial procedures $P_1, \ldots, P_n$.

1.2. Flowcharts of protocols. In this subsection we define how to transform instructions and concurrent protocols into flowcharts. A flowchart is a labeled directed graph with specified (and distinct) start and final nodes. If X is a flowchart then the set of nodes of X is denoted *locations*(X). The directed edges are called arrows. The only labels on arrows are the symbols T and F (for *true* and *false*) The labels of the nodes of X are among the following.

(1) "Start", and "Final". Nodes with these labels are called start and final nodes. Graphically, the *Start* label is a short arrow pointing towards the node, and the *Final* label is an oval.

(2) An assignment instruction of the form "$x := \tau$". Such nodes are called assignment nodes.

(3) An external instruction of the form "$R(t_0, \ldots, t_{n-1})$". Such nodes are called external instruction nodes.

(4) A condition (that is, a formula τ in language L) can appear as a label. (Such nodes are associated with either while or conditional instructions). Two arrows emanate from such nodes, and they are labeled with T (for true) and F (for false). Such nodes are called conditional nodes.

(5) The label "return" tags return nodes (these are used in "while" flowcharts).

(6) The start node of a combined flowchart is labeled with "concurrently do".

To each serial instruction s we define a serial flowchart X_s (see Figure 3.3 and below). It turns out that X_s has the following properties:

(1) There are distinct start and final nodes.

(2) A start node has (in addition to the start label) another label; a final node has no tag other than "final".

(3) Any nodes other than the start node have just one tag.

(4) A single arrow emanates from an assignment node (called an assignment arrow). A single arrow emanates from an external operation node (called an external operation arrow). A single arrow emanates from a return node (called a return arrow).

(5) Two arrows emanate from a conditional node: T and F arrows (leading to distinct nodes).

DEFINITION 1.2 (SERIAL FLOWCHARTS). *The definition of the serial flowchart* X_s *is by induction on the complexity of the serial instruction* s.

---

[1] This is a strong requirement, clearly much stronger than the interference freedom of Owicki and Gries [24]. What saves our definition from being only trivially applied to procedures operating in isolation is the fact that procedures *can* communicate via external operations.

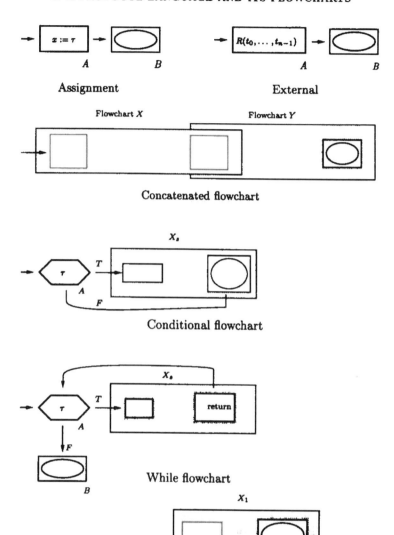

FIGURE 3.3. Flowcharts.

(1) To an assignment instruction s of the form $x := \tau$ we attach a flowchart X_s with two nodes A and B and an arrow from A to B. Node A is labeled with the assignment instruction $x := \tau$, it is the start node, and B is the final node (so "start" is an additional label on A, and "final" is a label on B). This flowchart is called an *assignment* flowchart.

(2) To an external instruction we attach again a two-node flowchart, but now the start node is labeled with the external instruction. This flowchart is called an *external instruction* flowchart.

(3) Suppose that $X = X_{s_1}$ and $Y = Y_{s_2}$ are the flowcharts attached to instructions s_1 and s_2 and assume that the sets of nodes *locations*(X) and *locations*(Y) are disjoint (this causes no problems of course because one can always find isomorphic, disjoint copies of the flowcharts). To the concatenated instruction $(s_1; s_2)$ we attach the following *concatenated flowchart* $X; Y$. It is obtained by identifying the final node of X with the start node S of Y. The resulting node is no longer a start/final node and its label is the non start label of S. The start node of the concatenated flowchart is the start node of X, and its final node is the final node of Y. There are no new arrows in $X; Y$, only X arrows and Y arrows.

(4) To a conditional instruction **if** τ **then** s we attach the following *conditional* flowchart Y, obtained from the flowchart X_s as follows. The start node of Y is a new node, A, labeled with condition τ ("new" means not in X_s). Two arrows emanate from A: one is labeled with "T" and it leads to the start node of X_s, and the other is labeled with "F" and it leads to the final node of X_s. This final node is taken to be the final node of the conditional flowchart Y as well.

(5) To a loop instruction **while** τ **do** s we attach the following *while* flowchart Y, constructed from X_s as follows. The start node of Y is a new node A labeled with condition τ, and the final node of Y is a new node B. Two arrows emanate from A: one labeled with "T" leads to the start node of X_s (which is no longer a start node), and the other labeled with "F" leads to B. An additional arrow (a "returning" arrow) is drawn from the final node of X_s (now labeled with "return") back to A.

This ends the definition of serial flowcharts. For each serial flowchart X_s we let $\mathcal{V}(X_s) = \mathcal{V}(s)$ be the set of variables that appear in the terms and formulas that tag X. We shall define now combined flowcharts for concurrent protocols.

Combined flowcharts. Suppose that a concurrent protocol P of the form

$$\textbf{concurrently} \;\; \textbf{do} \; P_1, \ldots, P_n \; \textbf{od}$$

is given, where the P_i are pairwise non-interfering serial procedures. Each procedure P_i is the name of an instruction s_i, and we let X_i be the serial flowchart attached to s_i. We may assume that the nodes of X_i, $1 \leq i \leq n$, are pairwise disjoint, and then the *concurrently combined flowchart* $X = X_P$ is defined as follows:

(1) Find a new start node for X (denoted $start_X$) and label it with

$$\textbf{concurrently} \;\; \textbf{do.}$$

(2) Draw "splitting" arrows from that new start node of X to each of the start nodes of the flowcharts $X_1, \ldots, X_n$.

We define $\mathcal{V}(X_P) = \bigcup_{1 \le i \le n} \mathcal{V}(X_i)$: the set of variables of X_P is a disjoint union of the sets of variables of its serial flowcharts. Note that a combined flowchart has n final nodes.

1.3. States and transitions. Recall that we have a fixed first-order language L with its variables V_L, and an interpretation B for L. Let $C \notin V_L$ be a new "location" variable. A state is a function assigning to a set of L-variables some values and to C a node in some flowchart. If S is a state then $\mathcal{V}(S)$ denotes the set of L-variables over which S is defined, so that $\{C\} \cup \mathcal{V}(S)$ is the domain of S. More specifically we have this

DEFINITION 1.3 (STATES). *Let X be a serial flowchart and let $C \notin V_L$ be a new "location" variable. S is a state over X (or an X-state) if*

(1) *$\mathcal{V}(X) = \mathcal{V}(S)$, that is S is defined over all variables of X. Let $V_0 = \mathcal{V}(S)$.*
(2) *$S|V_0$ is an assignment for B. That is, for every variable $v \in V_0$, $S(v)$ is an individual in B of the sort of v.*
(3) *$S(C)$ is a node in X; that is $S(C) \in locations(X)$.*

The states just defined are "regular" states, and there are also special types of states over X, called "unreachable" or "undefined" states which are used to express the fact that an operation is non-terminating. An unreachable state u is only defined on control variable C and is thus associated with a particular location $u(C) \in locations(X)$. So $\mathcal{V}(u) = \emptyset$, and $dom(u) = \{C\}$ for an unreachable state u. The collection of all states over X is denoted states(X).

Now let X be a concurrently combined flowchart obtained from pairwise non-interfering serial flowcharts $X_1, \ldots, X_n$. So if $V_i = \mathcal{V}(X_i)$ then $V = \bigcup_{1 \le i \le n} V_i$ is a disjoint union. A state S is said to be an X-state if either

(1) $\mathcal{V}(S) = V$ and $S(C) = start_X$ (and then S is called a global state), or
(2) for some i, S is an X_i-state.

For any flowchart X, serial or combined, we denote by *states*(X) the collection of all X-states.

For every combined flowchart X we assume a given set denoted *initial*(X) of initial, global states. Usually the initial states consist of *all* global states, but it is legitimate to restrict them further. We next define transitions; Section 4 contains a detailed example of this definition.

DEFINITION 1.4. *Let X be a serial flowchart. A pair $\langle S_0, S_1 \rangle$ of X-states is called a transition (over X). (We write interchangeably (X, Y) and $\langle X, Y \rangle$ to represent pairs—the difference being just a matter of graphical preference.) Transitions are used to represent executions of instructions. We define for every arrow α of X a collection $\mathcal{T}r_X(\alpha)$ of transitions over X. When the identity of X is evident, we write simply $\mathcal{T}r(\alpha)$. The definition of $\mathcal{T}r(\alpha)$ is given below, but we can say intuitively that a transition in $\mathcal{T}r(\alpha)$ represents an execution of α.*

Every arrow α is a pair (n_0, n_1) of nodes in X, and if $t \in \mathcal{T}r(\alpha)$ then t has the form $t = (S_0, S_1)$, where S_0 and S_1 are states over X such that $S_0(C) = n_0$

and $S_1(C) = n_1$. S_0 is always a regular state. If S_1 is also a regular state then t is said to be a terminating transition, and if S_1 is an unreachable state then t is said to be a non-terminating transition. In case t is terminating then $\mathcal{V}(S_1) = \mathcal{V}(S_2)$ always holds. We first define Tr_X when X is a serial flowchart.

(1) If $\alpha = (A, B)$ is an assignment arrow, then node A is labeled with an assignment instruction $x := \tau$, and we define $Tr(\alpha)$ as the set of all (terminating) X-transitions (S_0, S_1) such that:
   - $S_0(C) = A$ and $S_1(C) = B$.
   - For every variable $z \in \mathcal{V}(S_0) = \mathcal{V}(S_1)$: If $z \neq x$ then $S_0(z) = S_1(z)$, but $S_1(x)$ is the value of τ in B when evaluated with the assignment S_0.

(2) If $\alpha = (A, B)$ is an external operation arrow where A is labeled with $R(t_0, \ldots, t_{n-1})$, then we define $Tr(\alpha)$ to be the set of all X-transitions (S_0, S_1) such that
   - $S_0(C) = A$ and $S_1(C) = B$,
   - Either S_1 is the unreachable state with $S_1(C) = B$, or else $\mathcal{V}(S_0) = \mathcal{V}(S_1)$ and:
     for every variable $z \in \mathcal{V}(X)$: If z is some t_i in a *variable* place, then $S_1(z)$ can have any value (of the right type), but otherwise $S_0(z) = S_1(z)$.

   To say this less formally, there are two types of transitions related to an external instruction: non-terminating transitions and terminating. If (S_0, S_1) is a non-terminating transition, reflecting the case when the call to R has never returned, then S_1 is the unreachable state with location B. If (S_0, S_1) is a terminating transition, then only variables that are in a variable place may change values.

(3) Let A be a conditional node labeled with condition τ (a formula in L) corresponding either to a conditional or to a while instruction. Two arrows emanate from A: one labeled with T and the other with F. Say $t_1 = (A, B_1)$ is labeled with T, and $t_2 = (A, B_2)$ with F. Let T_0 be the set of all terminating transitions (S_0, S_1) over X such that $S_0(x) = S_1(x)$ for every L-variable x.
   - To the "True" arrow, t_1, Tr assigns all transitions in T_0 such that $S_0(C) = A$, $S_1(C) = B_1$ and formula τ is true in B under assignment S_0.
   - To the "False" arrow, Tr assigns all transitions in T_0 such that $S_0(C) = A$, $S_1(C) = B_2$, and τ is false in B under assignment S_0.

(4) To a returning arrow going from node n_0 to n_1, Tr assigns all transitions (S_0, S_1) in T_0 where $S_0(C) = n_0$ and $S_1(C) = n_1$.

This defines Tr_X for any serial flowchart. Now let X be a combined flowchart obtained from mutually non-interfering serial flowcharts $X_1, \ldots, X_n$. Let $start_X$ be the new start node of X. $\mathcal{V}(X) = \bigcup_{1 \leq i \leq n} \mathcal{V}(X_i)$ is a union of pairwise disjoint sets of variables.

To a splitting arrow α leading from $start_X$ to the start node of X_i there corresponds the set $Tr(\alpha)$ of all splitting transitions defined as pairs (S_0, S_1) of states such that

(1) $S_0(C) = start_X$ and $S_i(C)$ is the start node of X_i,
(2) $\mathcal{V}(S_0) = \mathcal{V}(X)$ and $\mathcal{V}(S_i) = \mathcal{V}(X_i)$,
(3) $S_1 \lceil \mathcal{V}(X_i) = S_0 \lceil \mathcal{V}(X_i)$.

In plain words, a splitting transition takes a global state over the combined flowchart (presumably in $initial(X)$) and restricts it to the variables of some X_i, thus serving as an initial state for X_i.

If α is an arrow in one of the flowcharts X_i, then $Tr_X(\alpha)$ is $Tr_{X_i}(\alpha)$ as defined above for the serial flowchart X_i. The set of all transitions attached to the arrows of a flowchart X is denoted $Tr(X)$. So $Tr(X) = \bigcup \{ Tr(\alpha) \mid \alpha$ an arrow of $X \}$.

2. Semantics of flowcharts

The semantics of flowcharts is described here in two phases. In the first, we associate with each flowchart its histories, which give the non-restricted semantics. Then restricted executions of these histories are defined which incorporate the specifications of the communication devices and yield the restricted semantics.

2.1. Histories and executions of flowcharts. Let X be a flowchart and $A = states(X)$ be its set of states. We want to define *histories* for X. For a single serial flowchart, a history is essentially a sequence of states with an index set that can either be N or a finite initial segment of N (we say essentially because a history, as we see it, is in fact a structure, not a sequence). When X is a combined flowchart with k serial flowcharts, then we have k independent processes executing concurrently and we therefore have k sequences rather than a single one.

A history for a serial flowchart X is a sequence of states σ, finite or infinite, such that the first state $\sigma(1)$ is an initial X-state and $\langle \sigma(i), \sigma(i+1) \rangle$ is a transition over X. If X is now a combined flowchart formed from k serial flowcharts $X_1, \dots, X_k$ then a history for X is a vector of sequence $H = \langle \sigma_1, \dots, \sigma_k \rangle$ such that:

(1) $\sigma_1(0) = \sigma_2(0) = \cdots = \sigma_k(0) = S_0$ is a global X-state.
(2) $\langle \sigma_i(0), \sigma_i(1) \rangle$ is a splitting transition with $\sigma_i(1)$ being the start node of X_i.
(3) $\sigma_i(1), \sigma_i(2), \dots$ is a history for X_i.

The problem with this definition is that H is not a structure capable of first-order definitions. We therefore redo the definition, but now aiming to obtain structures.

A sequence is a two-sorted structure of the form $(H, A; <, \sigma)$ where H is the sort of indices, interpreted either as the standard set N of natural numbers or as a finite initial segment of N, sort A is a set of possible values, $<$ is the natural ordering of H, and σ is a function from H into A. (Namely, σ is a sequence in the usual sense.) In case the domain of σ is the finite initial segment $[0, \dots m-1]$ we say that the sequence has length m. In case the domain of σ is N we say that the length of σ is ω (or just that the sequence is infinite). It is also assumed (as usual) that the structure is equiped with ordered pairs, and it is convenient (but not strictly necessary) to also have the successor function $+1$ in this structure.

The ordered pairs are used when A is the set of all states and we want to express properties of transitions, which are pairs of successive states in a state sequence.

To deal with k independent sequences we have a k-folded sequence structure of the form $S = (\mathbb{N}, A; <, \sigma_1, \ldots, \sigma_k)$, where each σ_i is a sequence of length d_i which is either ω or some natural number. The structure S is also equiped with ordered pairs, which allows us to define "steps". The "nth step by process i", for $1 \leq i \leq k$ and for n such that both n and $n+1$ are in the domain of σ_i, is the pair $s = (i, n)$. We say that step $s = (i, n)$ *represents* transition $t = \langle \sigma_i(n), \sigma_i(n+1) \rangle$, or that t is the transition corresponding to step s. In fact, some authors take histories as sequences of transitions rather than of states, and intuitively this makes sense because transitions are the substance of change. However, notationally, it seems easier to deal with sequences of states. (The two approaches are clearly equivalent.) Thus $(i, 0)$ is the first step, $(i, 1)$ the second step by process i, etc.

The steps are partially ordered: steps of different processes are incomparable, but the nth step of process i precedes the n'th step if $n < n'$. That is, we define $<_{steps}$ on the set of steps by $\langle i, n \rangle <_{steps} \langle i', n' \rangle$ iff $i = i'$ and $n < n'$.

Suppose a combined flowchart X is built from k serial flowcharts $X_1, \ldots, X_k$ that do not interfere with each other; let $start_X$ be the start node of X, and let $A = states(X)$ be the set of all X-states. Suppose that Tr has been defined on the arrows of X as in Definition 1.4.

DEFINITION 2.1 (HISTORY). *A history of a combined flowchart X (combined from k non-interfering flowcharts $X_1, \ldots, X_k$) is a k-folded sequence*

$$H = (\mathbb{N}, A; regular, <, \sigma_1, \ldots, \sigma_k)$$

such that $A = states(X)$, regular is a predicate on A (defining the set of regular states), and:

(1) *There is an initial, global X-state, S, such that $S = \sigma_i(0)$ for all $1 \leq i \leq k$. We say that S is the initial state of H.*

(2) *For every step (i, n) in H, $\langle \sigma_i(n), \sigma_i(n+1) \rangle \in Tr(X)$. That is, $\langle \sigma_i(0), \sigma_i(1) \rangle$ is a splitting transition, and $\langle \sigma_i(n), \sigma_i(n+1) \rangle$ is in $Tr(X_i)$ (whenever n, $n+1$ are in the domain of σ_i).*

We say that step $s = (i, n)$ is terminating iff $\langle \sigma_i(n), \sigma_i(n+1) \rangle$ is a terminating transition, that is $\sigma_i(n+1)$ is a regular step.

The collection of all possible histories of X is denoted $\mathcal{H}(X)$. This collection is defined to be the internal semantics of X. Neither "time" nor the precedence ordering on the events is represented in the internal semantics. The specifications of the diverse communication devices are not represented in $\mathcal{H}(X)$. Thus $\mathcal{H}(X)$ represents the unrestricted semantics.

A history is still a composite structure because its sort of states A is not an abstract set, but rather a specific structured set, namely the set of states of X. The flattening of H is called a *history model*. (See Chapter 1 for a definition of composite structures and flattenings.) In the following section we give more information on this particular flattening procedure. We shall continue to use $\mathcal{H}(X)$ to denote the collection of all history models for X.

DEFINITION 2.2 (EXECUTIONS OF HISTORIES). *Let H be a history model for a combined flowchart X as defined above. We say that a system execution $\mathcal{E}$ is a non-restricted execution of H if $\mathcal{E}$ is an expansion of H to a larger signature containing in addition a sort of events E, a partial ordering $\prec$ on E, and a predicate Terminating over E such that:*

(1) The set of events of $\mathcal{E}$, E, is the set of steps of H.
(2) For any events e_1, e_2 in E, if $e_1 <_{steps} e_2$ in H then $e_1 \prec e_2$.
(3) Any event $e \in E$ is terminating iff it is a terminating step. That is $Terminating(e)$ iff $e = (i, n)$ and $\sigma_i(n+1)$ is a regular state.

We let $\Sigma(X)$ be the set of all executions of histories of X.

Informally speaking, an execution of a history is any determination of a precedence relation on the events (steps) that respects the $<_{steps}$ ordering (and satisfies the general requirements made in the definition of a system execution, namely the Wiener–Russell property and the finiteness condition). The definition given here of history models and history executions is not completely satisfactory, because it leaves to the reader the work of precisely defining the flattening of the composite histories. We shall give a fuller definition in the following section, but before doing this we want to complete the definition of the external semantics, in terms of history executions.

2.2. External semantics. The external (restricted) semantics of a flowchart can be defined once the semantics of its communication devices (and external operations) is determined. The restricted semantics of X will be defined as a sub-collection of $\Sigma(X)$, consisting of all system executions of histories of X in which the communication devices operate properly. We have defined in Chapter 2 Definition 4.4 what constitutes a formal specification of a device. In short, a communication device D and its related external operations $R_1, \ldots, R_m$ are specified by a system S_D (that is a collection of system executions in some specific signature L_D). S_D contains all system executions in which the events pertaining to the operations $R_1, \ldots R_m$ describe a correct functioning of the device.

Now let X be a combined flowchart and let $D_1, \ldots, D_n$ be the devices used by X. Suppose that system S_{D_i} specifies device D_i for every i. Then the restricted semantics of X is the system S defined by

$$S = \Sigma(X) \| S_{D_1} \| S_{D_2} \cdots \| S_{D_n}$$

That is, following the definition of $\|$, a system execution is in S iff its reduct to the signature of $\Sigma(X)$ is in $\Sigma(X)$, and its reduct to the signature of each S_{D_i} is in S_{D_i}.

3. Histories as structures

Since the theme of this book is that structures in the Tarskian sense are the proper means to link concurrent programs with their executions, I cannot be satisfied with composite structures, and I must show how to transform them into structures that support first-order proofs and definitions. The point is this: according to its definition, the universe of a structure consists of abstract "points" with no inner composition, and the elements get their meanings only through the interpretation of the functions and predicates. Hence isomorphic structures are regarded as the same. The definition of histories deviates from this principle because of the requirement that $H = (\mathbb{N}, A, <, \sigma_1, \sigma_2 \ldots)$ is a history of X if $A = states(X)$. The points of sort A are thus required to have an inner composition, namely to be the states of X. In the terminology of Chapter 1, histories are composite structures, and since it is important (for correctness proofs) to be able to say exactly when a statement holds in a history, we must redefine histories as first-order structures, that is to flatten them.

For example, suppose that we want to express the statement that $\sigma_3(5)$ is a state that gives the value a to variable x. We soon realize that there is no formula in the first-order language for H that achieves this, and we must therefore expand H. In a sense, it is trivial to transform any mathematical object into a first-order structure, but in another sense it is a delicate matter requiring much thought, especially for complex objects, because there are different ways of achieving this, and some are more natural than others. For example, we may add a function symbol Val_x for every variable $x \in \mathcal{V}(X)$, with the intention that $Val_x(s)$ is the value of variable x at state s. Then the above required formula would be $Val_x(\sigma_3(5)) = a$. There are other possibilities though (see the discussion on page 8) which allow a more natural expression such as $\sigma_3(5)(x) = a$. Should we fix once and for all a particular signature and definition of histories as structures, or leave the minute details to the reader and only say what is expected from such structures? I prefer the second alternative, and for this the notion of definability will be very handy. It suffices for the above example to require that the relation $\sigma_i(n)(x) = a$ is definable where i is a process index, $n < d_i$ is a state index, x is a variable, and a is a value. In this way the definition of history models can be given without too many arbitrary details. Very roughly, one can say that a history model (for a history H) is a structure in which everything that we can say naturally about H is definable. But we must be a little bit more specific.

So let $\mathcal{B}$ be the background structure for our protocol language, X a combined flowchart, and H a history for X as a composite structure. A *history model* for H is a structure $\mathcal{S}$ for a rich enough signature such that:

(1) X is a reduct of $\mathcal{H}$. That is, $locations(X)$, the tags of X, and $states(X)$ are all sorts in $\mathcal{H}$. If $s \in states(X)$ and x is a variable, then the value $s(x)$ is definable in $\mathcal{H}$. There is a predicate indicating which states are regular and which are unreachable. The function associating a node in $locations(X)$ to its label is definable in $\mathcal{H}$. The arrows of X (and their labels) are also definable in $\mathcal{H}$. $\mathcal{H}$ contains enough information to describe the labels. For example, if $R(t_0, \ldots, t_{k-1})$ is an external operation label, then the statement that node $N \in locations$ with this label is an R node is expressible in $\mathcal{H}$. Moreover, there is a way of defining in $\mathcal{H}$ the value of any term t_i at each of the states in $states(X)$.

(2) The transitions are definable in $\mathcal{H}$, and the function Tr giving to each arrow its transitions is definable. $\mathcal{H}$ distinguishes between terminating and nonterminating transitions.

(3) H is a reduct of $\mathcal{H}$. But now the set $A = states(X)$ is a flat set of points, acquiring their meaning only through the functions and predicates of $\mathcal{H}$.

(4) $\mathcal{H}$ is equiped with pairs, and the set of steps $steps(X)$ is a flat set of points. The ordering $<_{steps}$ is definable. The transition associated with any step s is definable from s. That is, if $s = (i, n)$ is the nth step by process i, then the transition $\langle \sigma_i(n), \sigma_i(n+1) \rangle$ is definable from s. We say that s "executes" the instruction that tags $\sigma_i(n)(C)$. For every operation symbol R there is a predicate R_event in the signature of $\mathcal{H}$, and for every step s that executes R, $R\_event(s)$ holds.

The steps represent executions of the instructions of X. Specifically, step $s = (i, n)$ is an execution of the instruction tagged at node $\sigma_i(n)(C)$. If transition $\langle \sigma_i(n), \sigma_i(n+1) \rangle$ is terminating, then we say that s is a terminating step, but otherwise, if $\sigma_i(n+1)$ is an unreachable state, then we say that s is a nonterminating step.

We now need a detailed example to clarify the definitions given in the last three sections.

4. The Pitcher/Catcher example

The Pitcher/Catcher protocol is given in Figure 3.4 as two named procedures that execute concurrently: *Pitcher* and *Catcher*. Our intention is to go back to the beginning of this chapter and to follow all the definitions and see how they apply to this example, thus finally obtaining a definition of its semantics.

The *background structure* $\mathcal{B}$ is defined first. Its language L contains:

(1) Sort *numbers* is interpreted in $\mathcal{B}$ as N, the set of natural numbers.[2]
(2) A binary predicate $<$, interpreted as the usual "less than" relation on N.
(3) A constant 0, interpreted as zero of course.

The variables of L (and hence of the protocol) are a, *received*, and b, which vary through *numbers*.

The *Pitcher/Catcher* protocol language is based on L. So "*received* $< b$" is a formula of L.

Three external operations appear in the protocol: *obtain*, *throw*, and *catch*. All are unary. In *throw(a)*, a is a value parameter ranging over *numbers*. In *obtain(a)* and *catch(b)*, a and b are variable parameters ranging over *numbers*. So *throw(a)* does not change the value of variable a, and *catch(b)* determines the value of b. That is to say, after executing *catch(b)* variable b can change its previous value and acquire any value. The specifications of the *throw/catch* operations (from Chapter 2, Section 4.1) will only be considered at the stage of defining the external semantics of this protocol.

There are no restrictions on the value assigned to a in *obtain(a)*; any number is possible. We assume that *obtain* is always terminating (but *throw/catch* may be non-terminating).

The structure of the procedures follows the definition given in subsection 1.1. *Pitcher* for example is a **while** instruction. The condition of this while loop is the sentence **true** which is always evaluated to true, of course, and hence the loop

---

[2] Should this sort be *numbers* or *number*? The singular form is adopted by PASCAL for example, but the plural is sometimes more natural. I have not made any consistent commitment on this matter.

procedure *Pitcher*	procedure *Catcher*
begin	**begin**
while true do	*received* := 0;
(1) *obtain*(a);	**while true do**
(2) *throw*(a)	(1) *catch*(b);
end	(2) **if** *received* < b
	then *received* := b
	end
protocol *Pitcher/Catcher*	
concurrently do *Pitcher, Catcher* **od**	

FIGURE 3.4. Pitcher/Catcher protocol.

is executed forever. The body of this loop is itself a concatenated instruction obtained by concatenating two external instructions, *obtain*(a) and *throw*(a).

Notice that *Pitcher* and *Catcher* have no common variables whatsoever, and hence satisfy the non-interference property. The concurrent protocol

concurrently do *Pitcher, Catcher* **od**

can be defined.

The next step is to transform this two procedure protocol into a protocol flowchart. The serial flowcharts $X_{Pitcher}$ and the combined *Pitcher/Catcher* flowchart appear in Figure 3.5. The *Pitcher/Catcher* flowchart X is a concurrent combination of two serial flowcharts $X_1 = X_{Pitcher}$ and $X_2 = X_{Catcher}$. The careful reader has surely noticed that no F arrow leaves the *true* nodes 1 and 7; this is just for graphical clarity. The set of variables is $\mathcal{V}(X) = \{a, received, b\}$.

Now we describe the states of the Pitcher flowchart and the states of the Catcher flowchart. In general, a state is a map from variables to values. These variables include program variables and a "location" variable C such that $C \notin \mathcal{V}(X)$.

There are two different collections of local states: *Pitcher*-states and *Catcher*-states, which can either be regular or unreachable states. S is a regular *Pitcher*-state iff S is a function that assigns some *Pitcher*'s location to C (i.e. in $\{1, 2, 3, 4, 5\}$), and to variable a some natural number (which is in B). Similarly, T is a regular *Catcher* state iff it is a function that assigns to C a *Catcher*'s location, and to variable *received* and b some natural numbers. Any function with values in the correct range is a state, and so there are infinitely many states.

In addition to these "regular" states we have special "unreachable" states. These are needed to represent external (throw/catch) operations that do not return (are non-terminating). An unreachable state is a function like any other state, but it is only defined on control variable C. If u is an unreachable state then $u(C)$ is a node (4 or 9), but u is not defined on the program variables.

Initial states are also defined. S is a *Pitcher*'s initial state if it is in *Pitcher*-states and $S(C) = 1$. ($S(a)$ may be any natural number.) Intuitively, this is a state in which *Pitcher* is about to execute its first instruction. Similarly, T is

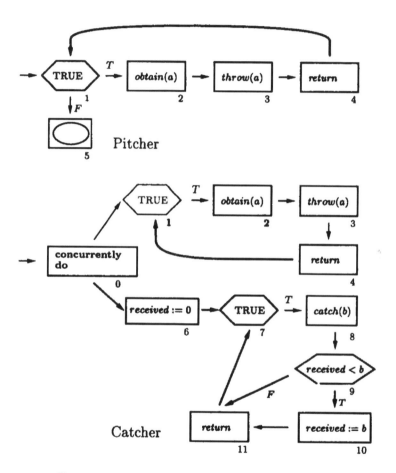

FIGURE 3.5. *Pitcher* and Combined *Pitcher/Catcher* flowcharts.

a *Catcher's* initial state if it is a *Catcher's* state that assigns to C the starting node 6.

Besides the local *Pitcher* and *Catcher* states, global states are also defined. S is a global state if $\mathcal{V}(S) = \mathcal{V}(X)$ and $S(C) = 0$.

We define transitions: these are pairs of states that are attached to the arrows of the flowchart and describe the passage from one node to another. There are two types of transitions: *terminating* and *non-terminating*. A transition (S_1, S_2) is terminating iff S_2 is a regular state. (S_1, S_2) is non-terminating iff S_2 is an unreachable state.

For every arrow α in the flowchart, the set of transitions attached to α is denoted $Tr(\alpha)$. Whenever the pair of states (S_1, S_2) is in $Tr(\alpha)$, and $\alpha = (X, Y)$, then $S_1(C) = X$ and $S_2(C) = Y$. Roughly stated, such a transition represents an execution of the instruction that labels node X, and it thus determines the relationship between the values of S_1 and S_2.

Consider the *Catcher* flowchart for example, to see how the *Catcher* transitions in $Tr(\alpha)$ are defined for $\alpha \in \{(6,7), (7,8), (8,9), (9,10), (9,11), (10,11)\}$.

(1) The $(6,7)$ arrow represents executions of the assignment instruction

$$received := 0.$$

To this arrow we attach the set of all transitions (T_0, T_1) where T_0 and T_1 are *Catcher*-states such that: $T_0(C) = 6$, $T_1(C) = 7$, $T_1(received) = 0$ and $T_0(b) = T_1(b)$.

(2) To the $(7,8)$ arrow leaving the **true** node we attach the set of all transitions (T_0, T_1) such that $T_0(C) = 7$ and $T_1(C) = 8$ but for each of the other variables (*received* and b) T_0 and T_1 give the same value. (Formally, there should also be an F arrow leaving node 6, but for clarity it is omitted from Figure 3.5.)

(3) An execution of $catch(b)$ may be terminating or non-terminating. A terminating execution may change the value of b, and any value (in its type) is possible. A non-terminating execution never returns and is not associated with any value. Thus $Tr(\langle 8, 9 \rangle)$ consists of two types of transitions. Non-terminating transitions are pairs (T_0, u) such that $T_0(C) = 8$, $u(C) = 9$, T_0 is a regular state, and u is an unreachable state. Terminating transitions (T_0, T_1) in $Tr(\langle 8, 9 \rangle)$ are such that $T_0(C) = 8$, $T_1(C) = 9$, and $T_0(received) = T_1(received)$. (Since no restriction on the values of b is made, any pair of values is accepted for $T_0(b)$ and $T_1(b)$. This reflects the assumption that at this stage any value for b is possible. Only the external semantics can limit these values.)

(4) The label at node 9 is the formula $received < b$. Any *Catcher* state is also an assignment in the model theoretic sense: a mapping from the variables to B. If S is a state then the "assignment of S" is just the function S restricted to the variables of the background structure B, i.e. to all variables except the control variable C. Thus, for example, the formula of node 9 holds for the assignment of a state S exactly if $S(received) < S(b)$. A transition (T_0, T_1) is in $Tr(\langle 9, 10 \rangle)$ iff $T_0(C) = 9$, $T_1(C) = 10$, and formula $received < b$ holds for the assignment of S_0 (in the background structure).

A transition (T_0, T_1) corresponds to the "False" arrow iff $T_0(C) = 9$, $T_1(C) = 11$, and the formula is false under the assignment of T_0. In both cases T_1 does not change the program variables *receive* and b.

(5) To arrow $(10, 11)$ we attach the set of all *Catcher*'s transitions (T_0, T_1) such that $T_0(C) = 10$, $T_1(C) = 11$, $T_1(received) = T_0(b)$, and $T_1(b) = T_0(b)$.

(6) To the return arrow $(11, 7)$ we attach all terminating *Catcher*'s transitions that only change location from 11 to 7.

The set of transitions thus defined (corresponding to the arrows of the *Catcher* flowchart) is called "the *Catcher*'s set of transitions". In a similar manner transitions are attached to each arrow of the *Pitcher* flowchart, and they form "the *Pitcher*'s set of transitions".

Splitting transitions are also defined. $Tr(\langle 0, 1 \rangle)$ consists of all transitions (S_0, S_1) with $S_0(C) = 0$, $S_1(C) = 1$, and such that S_0 is a global state with $V(S_0) = \{a, received, b\}$ and S_1 is a *Pitcher* state with $S_1(a) = S_0(a)$. Similarly for the splitting arrow $\alpha = (0, 6)$ we define $(S_1, S_2) \in Tr(\alpha)$ iff S_1 is a global state and S_1 is the regular *Catcher* state defined by $S_2(C) = 6$ and $S_2|\{received, b\} = S_1|\{received, b\}$. Observe that the transitions tell us nothing about the connection between the *throw* and the *catch* operations. This is why they form the basis of the "unrestricted" semantics. Any possible behavior, even malfunctioning, can be represented by these transitions.

We continue with Section 2 to define histories for the combined flowchart X, and then executions of histories—non restricted and restricted—that define the semantics of X, the *Pitcher/Catcher* flowchart.

A history for X is a two-folded sequence $H = (N, A; <, \sigma_1, \sigma_2)$ where $A = states(X)$ is the collection of all states of the *Pitcher/Catcher* flowchart, $<$ is the ordering of the natural numbers N, and $\sigma_1 : H_1 \to A$, $\sigma_2 : H_2 \to A$ are two sequences of states, where H_i for $i = 1, 2$ is either N or a finite initial segment $[0, \ldots, d_i)$. There are two requirements that must be satisfied:

(1) $\sigma_1(0) = \sigma_2(0)$ is a global X-state. For $n > 0$ $\sigma_1(n)$ (when defined) is a *Pitcher* state, and $\sigma_2(n)$ (when defined) is a *Catcher* state.

(2) For every n such that $n + 1$ is in the domain of σ_1 (or σ_2) $\langle \sigma_1(n), \sigma_1(n+1) \rangle$ (or respectively $\langle \sigma_2(n), \sigma_2(n+1) \rangle$) is a transition in $Tr(X)$.

So $\langle \sigma_i(n) \mid n < d_i \rangle$ (or $\langle \sigma_i(n) \mid n \in N \rangle$ is a sequence of states that start with a splitting transition and is followed by *Pitcher* transitions for $i = 1$ and *Catcher* transitions for $i = 2$.

It is convenient to speak about *steps*. A *Pitcher* step is a pair $(1, n)$ such that $n + 1$ (and hence n) is in the domain of σ_1, and a *Catcher* step is a pair $(2, n)$ such that $n + 1$ is in the domain of σ_2. Corresponding to step (i, n) (where $i = 1, 2$) we have the transition $\langle \sigma_i(n), \sigma_i(n + 1) \rangle$. (Intuitively then, a step represents an execution of a transition.) A partial ordering $<_{steps}$ is defined on the steps by $\langle i_1, n_1 \rangle <_{steps} \langle i_2, n_2 \rangle$ iff $i_1 = i_2$ and $n_1 < n_2$. Namely *Pitcher* steps and *Catcher* steps are incomparable, and within each process the steps are ordered linearly.

Histories do not express any temporal information on the steps. Relation $<_{steps}$ only orders the steps of each process, but not steps from different processes, and $<_{steps}$ expresses not so much the temporal precedence relation as the logical interpretation of what constitutes an execution of a serial flowchart.

Therefore we must expand histories and relate them with some temporal information if we want to model reality. This is the role of executions of histories. Informally, an execution of a history is a system execution in which the steps are called events and a precedence relation $\prec$ is defined on them which extends the $<_{steps}$ ordering. The problem with this simple definition is that if system executions are to be regarded as expansions of histories, then histories must themselves be first-order structures. Yet in our definition above $H = (\mathbb{N}, A, <, \sigma_1, \sigma_2)$ is not a first-order structure, since A is required to be a very special set—namely *states*(X). This is one reason why histories must be flattened into history structures.

A deeper problem with this definition of histories (as composite structures) is this. We want a framework for proving correctness of protocols. As explained in Chapter 1 Section 2.2, any correctness condition that we want to prove will be written as a first-order sentence α. For an informal, mathematical proof that α holds in every history for X, the definition of histories as composite structures is good enough. However, for a fully formal proof conducted in first-order logic, one needs models as defined in Chapter 1. We shall therefore redefine histories as history models, and then the treatment of proving or disproving statements about histories will fall under a well known and general theory of logic.

Recall that X is the combined *Pitcher/Catcher* flowchart. Let

$$H = (\mathbb{N}, A; <, \sigma_1, \sigma_2)$$

be a history of X. If $\mathcal{H}$ is a flattening of H (as in Chapter 1 subsection 2.4) then we say that $\mathcal{H}$ is a *history model* for the *Pitcher/Catcher* flowchart. To be a little bit more precise we give some details on how $\mathcal{H}$ looks like.

(1) The main sort in $\mathcal{H}$ is the sort of steps $E = steps(H)$ which is now devoid of any inner composition. Any step $e \in E$ is a flat element of the universe of $\mathcal{H}$ (rather than a pair). There is a sort of "processes id" which is composed of 1 and 2 (or *Pitcher* and *Catcher*), and a sort of "numbers" (namely $\mathbb{N}$) and to any step s corresponds a pair (i, n). The functions $ProcId(e) = i$ and $StepNumber(e) = n$ give the inner composition of step e as a pair. Steps are the main sort because they represent events, but in isolation they give no information and the needed additional information is obtained from the other sorts, function, and predicates of $\mathcal{H}$.

(2) The background structure $\mathcal{B}$ is a reduct of $\mathcal{H}$. This just means that $\mathbb{N}$ ($= Catch\_Values$) is a sort in $\mathcal{H}$ and that $+, <, 0, 1$ are in $\mathcal{H}$ as well.

(3) The flowchart X is also a reduct of $\mathcal{H}$. This means that
   (a) *locations*$(X) = \{0, \ldots, 11\}$ is a sort of $\mathcal{H}$,
   (b) *arrows*$(X) = \{\langle 0, 1 \rangle, \ldots, \langle 11, 7 \rangle\}$ is a relation in $\mathcal{H}$,
   (c) $A = states(X)$ is a sort in $\mathcal{H}$.
   Again, A is represented as a flat set and additional features are required to reveal its character. For example, we may add a sort of "variables" $\mathcal{V}(X) = \{a, received, b\}$, and a function *apply* such that *apply*(s, x) gives the value of variable x at state s, this value being in the right sort of $\mathcal{B}$. It is more natural to write $s(x)$ instead of *apply*(s, x), if only we remember that s is an element of A rather than a function symbol. Similarly, $s(C) \in$

locations(X) gives the location of state s in one of the nodes. In addition, "regular" and "unreachable" are predicates defined on the states.

(4) The "terminating" predicate is such that a step (i, n) is terminating iff $\sigma_i(n + 1)$ is a regular state.

(5) Transitions are certain ordered pairs of states. The signature of $\mathcal{H}$ is rich enough so that $Tr(\alpha)$ is definable in $\mathcal{H}$. This means that there is a formula $\tau(i_1, i_2, s_1, s_2)$ with four variables such that for every arrow (i_1, i_2) in X and states $s_1, s_2 \in A$, $\tau(i_1, i_2, s_1, s_2)$ holds exactly when $\langle s_1, s_2 \rangle \in Tr(\langle i_1, i_2 \rangle)$. The reader may pause here and try to write down— if not all of τ then enough to be convinced that a program can be written which takes X as an input and produces τ (the syntactical specifications of the external operations are also needed).

(6) $\sigma_1, \sigma_2 : \mathbb{N} \to A$ are functions in $\mathcal{H}$. The requirement on σ_i is that they form a history, i.e., that $\sigma_1(0) = \sigma_2(0)$ is a global state and $\langle \sigma_i(n), \sigma_i(n + 1) \rangle \in Tr(X)$ for every n (in its domain). This requirement can now be formulated as a first-order statement: namely that for every n, if $s_1 = \sigma_i(n)$, $s_2 = \sigma_i(n + 1)$, $i_1 = s_1(C)$, $i_2 = s_2(C)$, then $\tau(i_1, i_2, s_1, s_2)$.

(7) Predicates *obtain_event*, *throw_event*, and *catch_event* are defined on the steps. Specifically, *obtain_event*($\langle i, n \rangle$) if and only if $\sigma_i(n)(C) = 2$ (and $i = 1$). *throw_event*($\langle i, n \rangle$) if and only if $\sigma_i(n)(C) = 3$ (and $i = 1$), and *catch_event*($\langle i, n \rangle$) if and only if $\sigma_i(n) = 8$ (and $i = 2$).

For any terminating catch/throw step $e = (i, n)$, $Value(e)$ is defined as follows. In general, the value of any step (i, n) that corresponds to an external r-place operation R is defined as the r-tuple of values of the parameters of R in $\sigma_i(n + 1)$, which implies that $Value(s)$ is not defined for non-terminating steps (whose second state is unreachable). In our protocol we get the following.

If $s = (2, n)$ is a terminating *catch_event* step, then

$$Value(s) = \sigma_2(n + 1)(b).$$

So $Value(s) \in Catch\_Values$ $(= \mathbb{N})$ in this case.
In case $s = (1, n)$ is a terminating *throw_event*, then

$$Value(s) = \sigma_1(n)(a).$$

Note that the value of a variable parameter (such as b) is taken in the second state (i.e., in $\sigma_i(n + 1)$), but the value of a value parameter (such as a) is taken in $\sigma_i(n)$. Since $\sigma_i(n)(a) = \sigma_i(n+1)(a)$, we could also define $Value(s) = \sigma_i(n + 1)(a)$.

Finally, if $s = (1, n)$ is an *obtain_event* (which we assume to be always terminating), then

$$Value(s) = \sigma_1(n + 1)(a).$$

From now on we write $H = (steps, \mathbb{N}; <, X, \sigma_1, \sigma_2)$ for a history model, even though this short notation does not include all objects that should be in such a structure.

If H is a history model of the *Pitcher/Catcher* protocol, then an execution of H is a system execution S that is obtained as an expansion of H with a partial ordering $\prec$ defined on the steps, which are now called the events, such that $\prec$ extends $<_{steps}$.

An execution of a history for X is also called an execution of X. $\Sigma(X)$ denotes the system of all executions of X. This is the non-restricted system: in $\Sigma(X)$ we find all executions, even those that describe faulty operations of the throw/catch registers. If we restrict the executions to respect the specifications of Chapter 2 subsection 4.1, then the restricted or external semantics is defined. That is, the external semantics of the *Pitcher/Catcher* protocol is the system S defined by

$$S = \Sigma(X) \| S_{throw/catch}$$

where $S_{throw/catch}$ is the system of the *throw/catch* specification in Chapter 2 (page 41).

We can restate this directly by defining S as the system of all history models (that is all executions of the protocols) such that it is possible to define a function α on the terminating *catch* events, and such that $\alpha(c)$ is a throw event satisfying

$$\alpha(c) \prec c \text{ and } Value(\alpha(c)) = Value(c).$$

4
Correctness of protocols

In the previous chapter we have defined internal and external semantics of protocols, but we did not explain how to use this definition to verify them—which is the aim of our chapter. We shall use an example to illustrate how to prove properties of protocols (the mutual-exclusion protocol of Peterson, described in Chapter 1).

Given a concurrent protocol P (in language L), we have seen how to define its combined flowchart X, and the internal and external semantics of X when the specification of the communication devices is given. The external semantics of X is a collection S of system executions in a language containing both L and the language in which the communication devices were specified. Intuitively, S represents all possibles runs of X in which the specification devices work properly. By proving that a property τ (a sentence in L) is true about X, we mean proving that τ holds in every system execution in S. This gives a precise mathematical explication (which one may accept or reject) of the notion of protocol correctness. The question of how one actually *proves* that a sentence holds in every structure in S is quite different. It is not my aim to set up a formal theorem-proving system, and I will remain within the semi-formal habitat of mathematics in proving such correctness statements. Still, practice is needed and examples are required to substantiate a claim that any given semantics is not just a mathematical definition but a useful tool for the programmer who wants to prove properties of protocols.

We begin with a very simple example of a correctness proof, and the following chapters contain more elaborate protocols that use divers communication devices.

1. Mutual exclusion revisited

We return here to Peterson's critical section protocol described in Chapter 1 (Section 3) and to the proof of the mutual-exclusion property. We redo the proof, but now with the semantics of chapter 3 rather than the assertional semantics. To begin with, we must rewrite the protocol because several features of the protocol as presented in Chapter 1 do not suit our definition of a protocol's language.

procedure P_1	procedure P_2
begin	begin
repeat_forever	repeat_forever
(1) NonCS;	(1) NonCS;
(2) $Write_{Q_1}(\text{true})$;	(2) $Write_{Q_2}(\text{true})$;
(3) $Write_{Turn}(1)$;	(3) $Write_{Turn}(2)$;
(4) repeat	(4) repeat
$\quad Read_{Q_2}(r_1); Read_{Turn}(s_1)$	$\quad Read_{Q_1}(r_2); Read_{Turn}(s_2)$
$\quad$ until $r_1 = \text{false} \vee s_1 = 2$	$\quad$ until $r_2 = \text{false} \vee s_2 = 1$
(5) CS;	(5) CS;
(6) $Write_{Q_1}(\text{false})$	(6) $Write_{Q_2}(\text{false})$
end	end

FIGURE 4.6. Peterson's protocol

(For example, we were using assignment instructions such as $Q_1 := \text{true}$ which treats a register as though it were a local variable, whereas now we shall write it as $Write_{Q_1}(\text{true})$.) The new form of procedures P_1 and P_2 is given in Figure 4.6.

The background structure here is particularly simple. Its universe consists of four distinct individuals $\{1, 2\} \cup \{true, false\}$ that interpret the four constants 1, 2, true, and false in its signature. Individuals $\{true, false\}$ form a sort, called *boolean*.

Next we define eight external operations for the protocol language. These are $Write_{Q_1}$, $Read_{Q_1}$, $Write_{Q_2}$, $Read_{Q_2}$, $Write_{Turn}$, $Read_{Turn}$, $NonCS$, and CS. We assume for simplicity that these operations are terminating.

(1) $Write_{Q_1}$, $Write_{Q_2}$ are unary operations taking value parameters of type *boolean*.

(2) $Read_{Q_1}$, $Read_{Q_2}$ are unary operations taking variable parameters of type *boolean*.

(3) $Write_{Turn}$ and $Read_{Turn}$ are unary operations, the first takes a value parameter, and the second a variable parameter, of type $\{1, 2\}$.

(4) NonCS and CS are zero-place operations.

Registers Q_1, Q_2, and *Turn* are assumed to be serial. (Seriality of registers is defined on page 36). The only assumptions made on operations NonCS and CS is that they are terminating.

The reader may have noticed that instructions of the form repeat-until (and repeat forever) were used in the protocol, despite the fact that in our protocol language only while loops were defined. We propose to enlarge the protocol language definition, and to add the following line to the BNF definition scheme. (I believe that at the moment of defining the general protocol language it was a reasonable strategy to make it as simple as possible.)

⟨*repeat*⟩ $\longrightarrow$ repeat ⟨*instruction*⟩ until τ | repeat_forever ⟨*instruction*⟩,

where τ is a formula in L. To the flowchart construction scheme we add the following item. Suppose that X_s is the flowchart attached to instruction s.

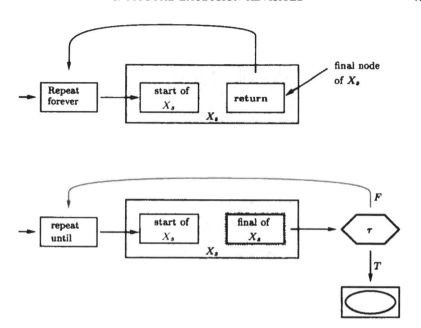

FIGURE 4.7. Flowcharts of repeat instructions

Then to repeat instructions

$$\textbf{repeat\_forever} \; s,$$
$$\textbf{repeat} \; s \; \textbf{until} \; \tau,$$

we attach flowcharts as depicted in Figure 4.7.

Procedures P_1 and P_2 do not interfere with each other since $\mathcal{V}(P_1) = \{r_1, s_1\}$ and $\mathcal{V}(P_2) = \{r_2, s_2\}$. The concurrent protocol (Peterson's protocol)

$$\textbf{concurrently} \;\; \textbf{do} \; P_1, P_2 \; \textbf{od}$$

can be formed. The combined flowchart X of this protocol is in Figure 4.8. So $\mathcal{V}(X) = \{r_1, s_1, r_2, s_2\}$.

To define X-states we add a location variable C.

(1) An initial global state is a function S defined on $\{C\} \cup \mathcal{V}(X)$ assigning to variables values in their sorts and such that $S(C) = start_X = 0$.

(2) P_1-states are functions S defined on $\{C\} \cup \{r_1, s_1\}$, with $S(C) \in \{1, \dots, 11\}$, $S(r_1)$ a boolean value, and $S(s_1) = 1, 2$.

(3) P_2-states are functions defined on $\{C\} \cup \{r_2, s_2\}$ with $S(C) \in \{12, \dots, 22\}$, $S(r_2)$ a boolean value and $S(s_2) \in \{1, 2\}$.

Since all operations are assumed to be terminating here, we have only regular states. For a more general discussion the reader may want to add unreachable states as well. If S is a state then $\mathcal{V}(S)$ denotes its program variables.

To each arrow α of X a set of transitions $\mathcal{T}r(\alpha)$ is defined. We give a few illustrative examples (compare with Definition 1.4 in Chapter 3).

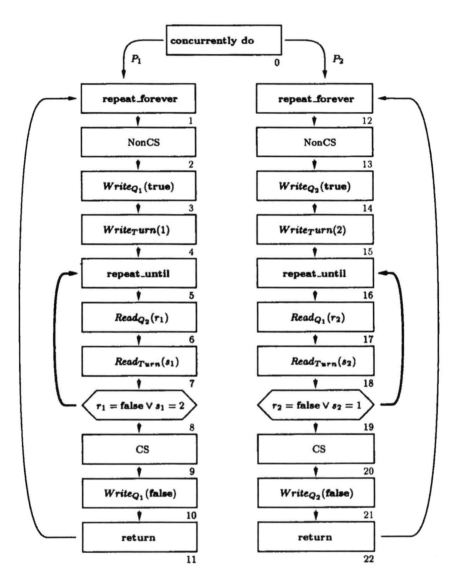

FIGURE 4.8. Combined flowchart

(1) For arrow $(0,1)$, which is a splitting arrow, $Tr(0,1)$ is the set of all transitions (S_0, S_1) such that S_0 is an initial global state, S_1 is a P_1-state, $S_1|\{r_1, s_1\} = S_2|\{r_1, s_1\}$ and $S_1(C) = 1$.

(2) For some arrows $\alpha = (n_1, n_2)$ the set of transitions $Tr(\alpha)$ consists of all pairs (S_1, S_2) that only change location from n_1 to n_2 and do nothing else. These are arrows $\alpha = (1,2)$, $(2,3)$, $(3,4)$, $(4,5)$, $(9,10)$, $(10,11)$, $(11,1)$. For example, $Tr(1,2)$ is the set of all transitions (pairs (S_0, S_1) of P_1-states) such that $S_1|\mathcal{V}(S_1) = S_0|\mathcal{V}(S_1)$ and $S_0(C) = 1$, $S_1(C) = 2$.

$Tr(2,3)$ is similarly defined, but if we also want to accept the possibility of non-terminating NonCS executions, then we should add to $Tr(2,3)$ transitions of the form (S_0, S_1) such that $S_0(C) = 2$, $S_1(C) = 3$ and S_1 is an *unreachable* state. (The reader should not be perplexed by our statement that transitions in $Tr(2,3)$ do not change the value of any variable: we deal here with the internal semantics of P_1 which refers only to the local variables.)

(3) For $\alpha = (6,7)$ define $(S_0, S_1) \in Tr(\alpha)$ iff S_0 and S_1 are P_1-states, $S_0(C) = 6$, $S_1(C) = 7$, and $S_0(s_1) = S_1(s_1)$. That is, S_1 may only change the value of r_1. $Tr(7,8)$ is defined similarly—its transitions may only change s_1. These are read transitions which can only change their read variables.

(4) $Tr(8,9)$ contains all P_1-transitions (S_0, S_1) with $S_0(C) = 8$, $S_1(C) = 9$, and such that $r_1 = false \lor s_1 = 2$ holds in the assignment S_0. $Tr(8,5)$ is defined with the negation of that statement instead.

A *history* for X is a structure $\langle N, states(X), regular, <, \sigma_1, \sigma_2 \rangle$ where σ_1, σ_2 are finite or infinite sequences of X-states such that $\sigma_1(0) = \sigma_2(0)$ is a global initial state and $\langle \sigma_i(n), \sigma_i(n+1) \rangle$ for $i = 1, 2$ are transitions. A history model is obtained by flattening a history for X. Let $\mathcal{H}$ be a history model for the combined flowchart X. Recall that a step s is a pair $\langle i, n \rangle$ (also written as (i, n)) such that $i = 1, 2$ is the process that executed that step (for P_1 and P_2) and n says that s is the nth step by P_i.

We are going to define predicates and functions over the steps in $\mathcal{H}$ that will be used in the correctness proof. Since these predicates and functions are *definable* in $\mathcal{H}$, the proof can be done in the language of $\mathcal{H}$ alone, replacing defined objects with their definitions. However, the point in having these definable predicates and functions is to make the proofs shorter and more natural.

(1) The predicates are: P_1, P_2, cs, $Write_{Q_1}$, $Write_{Q_2}$, $Write_{Turn}$, $Read_{Q_1}$, $Read_{Q_2}$, $Read_{Turn}$. (The reader should rely on the context to distinguish between instructions and predicates with the same name. So, for example, $Write_{Q_1}$ is an external operation symbol when it appears in the protocol, but it is a predicate in the discussion below.)

(2) The functions are: $Value$, q, t, c. These are partial functions on the steps.

The predicates are all defined on the steps of H, and their definition is as follows:

(1) $P_1(s)$ holds exactly if s is a P_1 step, that is, if $s = (1, n)$ for some n. $P_2(s)$ is similarly defined whenever s is a P_2 step.

(2) If $s = (i, n)$ is a step, then $cs(s)$ says that s is an execution of the critical section. Thus if $P_1(s)$ (or $P_2(s)$) then $cs(s)$ iff $location(\sigma_1(n)) = 9$ (or

$location(\sigma_1(n)) = 20)$.

(3) For any step $s = (i,n)$, $Write_{Q_1}(s)$ iff $i = 1$, and $location(\sigma_1(n)) = 3 \lor location(\sigma_1(n)) = 10$.

(4) All other predicates that describe executions of instructions are similarly defined.

We say that s is a *Write* step, denoted $Write(s)$, if $Write_{Q_1}(s) \lor Write_{Q_2}(s) \lor Write_{Turn}(s)$. Similarly $Read(s)$ is defined when s is a read step.

The function $Value$ is defined on the *Read/Write* steps. For example, if $s = (i,n)$ is a step such that $Write_{Q_1}(s)$ then $Value(s) = true$ iff $location(\sigma_1(n)) = 3$, and $Value(s) = false$ iff $location(\sigma_1(n)) = 10$. Less formally, if s is associated by σ to a terminating transition that describes an execution of a write instruction, then $Value(s)$ is the value written. Similarly, if s is a read step then $Value(s)$ is defined. For example, suppose that $s = (1,n)$, and $Read_{Turn}(s)$. This implies that $location(\sigma_1(n)) = 7$, and we define $Value(s) = \sigma_1(n+1)(s_1)$. Less formally, $Value(s)$ for a (terminating) read step s is the value of the receiving variable as found in the second state of the step.

We now define the functions d, t, and q on the CS steps. An informal definition is given first.

(1) Suppose that $s = (1,n)$ is a *cs* step in P_1 (for example). That is, $S(C) = 9$ for the state $S = \sigma_1(n)$. Then $(S(r_1) = false) \lor (S(s_1) = 2)$ follows, and hence either step $\langle 1, n-2 \rangle$ (the read of *Turn* into s_1) obtained 2, or $\langle 1, n-3 \rangle$ (the read of Q_2 into r_1) obtained *false*. Accordingly, we define $d(s) = \langle 1, n-2 \rangle$ or $d(s) = \langle 1, n-3 \rangle$. Even less formally, we can say that $d(s)$ for any *cs* step s is that read step that obtained the right condition for entering the critical section.

(2) Immediately before entering the 5–8 loop (or the 16–19 loop) the process had to write 1 (or 2) on *Turn*, and this write step is $t(s)$.

(3) Immediately before $t(s)$, the process had to write *true* on Q_1 (or on Q_2): this write is $q(s)$.

These definitions are somewhat informal. Formally, each of the functions q, t, d is defined by means of a formula $\phi(x,y)$ with two free step-variables x and y for which $\forall x \exists! y \phi(x,y)$ holds in $\mathcal{H}$[1]. Such formulas define functions that take any x to the unique y for which $\phi(x,y)$ holds.

The formal definition of t is given now, leaving the corresponding definitions of q and d to the reader. Let $\phi_1^t(x,y)$ be the formula

$$P_1(x) \land P_1(y) \land$$
$$Write_{Turn}(y) \land y <_{steps} x \land$$
$$\forall z \in steps \ [P_1(z) \land Write_{Turn}(z) \land z <_{steps} x \longrightarrow z \leq_{steps} y].$$

In words, this formula says that x and y are P_1-steps, and that y is the rightmost P_1-step before x that is a write on *Turn*.

We want to prove that for any step s in the history, if $P_1(s) \land cs(s)$ then there exists a single step w such that $\phi_1^t(s,w)$ holds in $\mathcal{H}$. This formula speaks only about steps of P_1, and its truth or falsity depends only on the P_1 part

---

[1]For any formula ψ, $\exists! y \psi(y)$ is a shorthand for $\exists y \psi(y) \land \forall y_1, y_2 \ (\psi(y_1) \land \psi(y_2) \to y_1 = y_2)$.

of $\mathcal{H}$, namely on σ_1. The proof that $\forall x \in P_1 \; \exists! y \in P_1 \; \phi_1^t(x, y)$ is therefore a proof about (a rather uninteresting) serial process, and it can be conducted in any of the formal methods developed for such processes. We argue in the usual mathematical fashion as follows.

We first prove for every cs step s in P_1 that there is a P_1 step $w <_{steps} s$ with $Write_{Turn}(w)$, and then we conclude that there is some ω such that $\phi_1^t(s, w)$. Now none of the first ten steps in P_1 (beginning with $\langle 1, 0 \rangle$) is a cs step. Hence (since the P_1 steps are linearly ordered) any cs step s in P_1 is after the fifth step which is a $Write_{Turn}$ step. Thus any cs step is preceded by a $Write_{Turn}$ step. Since the P_1 steps are linearly ordered, the finiteness condition for $<_{steps}$ implies that for every cs step s there is a maximal $Write_{Turn}$ step w in P_1 such that $w <_{steps} s$. Thus $\phi_1^t(s, t)$. Uniqueness of w is an obvious consequence of maximality. This formula defines the function t, namely for every cs step s in P_1, $\phi_1^t(s, t(s))$. In a similar manner ϕ_2^t is defined on P_2 cs steps.

The functions d and q are defined similarly.

The following lemma lists those crucial properties of the functions q, t, d that will be used in later stages of the proof. These properties are either properties of P_1-steps or of P_2 steps; no property involves both type of steps. Hence each of the sets of properties is about a serial process and can be proved by considering its serial flowchart only.

LEMMA 1.1. *For every step s by P_1, if $cs(s)$ then $q(s)$, $t(s)$, and $d(s)$ are defined and are such that:*

(1) $q(s) <_{steps} t(s) <_{steps} d(s) <_{steps} s$ *are all steps in P_1.*

(2) $Write_{Q_1}(q(s))$ *and* $Value(q(s)) = true$. *For every step w such that $Write_{Q_1}(w)$, either $w \leq_{steps} q(s)$ or else $s <_{steps} w$.*

(3) $Write_{Turn}(t(s))$ *and* $Value(t(s)) = 1$. *For every step v, if $Write_{Turn}(v)$ and $Value(v) = 1$, then either $v \leq_{steps} t(s)$ or else $s <_{steps} v$.*

(4) $d(s)$ *satisfies one of the two alternatives:*

    (a) $Read_{Q_2}(d(s)) \wedge Value(d(s)) = false$, *or*

    (b) $Read_{Turn}(d(s)) \wedge Value(d(s)) = 2$.

A similar statement is made for every step in P_2, interchanging 1 and 2.

Proof. I prove the third item as an example. By definition of $t(s)$, $\phi_1^t(s, t(s))$ holds. Hence $t(s)$ is a P_1 step such that $Write_{Turn}(t(s))$, and $t(s)$ is the $<_{step}$ last $Write_{Turn}$ step preceding s.

Clearly $Value(t(s))$ is 1, because any write on $Turn$ in P_1 has value 1. Now let v be any step such that $Write_{Turn}(v)$ and $Value(v) = 1$. It follows that v is a P_1 step, and hence $\phi_1^t(s, t(s))$ implies $v <_{step} s \rightarrow v \leq_{step} t(s)$. Using linearity of $<_{step}$ on the steps of P_1, this implies that either $v \leq_{step} t(s)$ or else $s <_{step} v$ ($s = v$ is not possible). $\square$

Let S be an execution of $\mathcal{H}$. So S is a system execution with set of events E that are the steps of $\mathcal{H}$, and such that $\prec$ is a precedence relation on the events which we assume to extend the step ordering. Since S is obtained by augmenting $<_{steps}$ to $\prec$ without any other alteration, the properties of Lemma 1.1 continue to hold in S with $\prec$ and $\preceq$ substituting $<_{steps}$ and $\leq_{steps}$ (this absoluteness is easily proved).

In order to prove that the protocol satisfies the mutual exclusion property, we must also assume that the registers are operating correctly—this is the external assumption which we now make. Let us further expand S by a return function ω and assume that the read/write events and the functions $Value$ and ω satisfy the requirements made on serial registers (in page 36).

We prove now the mutual exclusion statement MUTEX.

$$\forall c_1, c_2 \in E(cs(c_1) \wedge cs(c_2) \wedge c_1 \neq c_2 \rightarrow c_1 \prec c_2 \vee c_2 \prec c_1).$$

The proof relies only on the seriality of the registers and processes and on the properties enumerated in the lemma above. Since each of P_1 and P_2 are serial, two distinct critical section events that are both in P_1 (or both in P_2) are obviously $\prec$-comparable. Let c_1, c_2 be two critical-section events with $P_1(c_1) \wedge P_2(c_2)$, and we will prove that $c_1 \prec c_2 \vee c_2 \prec c_1$. The key point is to compare $t(c_1)$ with $t(c_2)$, the two last writes on $Turn$ preceding c_1 and c_2 in P_1 and P_2 respectively. It follows that both are write events on $Turn$ with $Value(t(c_1)) = 1$ and $Value(t(c_2)) = 2$. So (since $Turn$ is serial) either $t(c_1) \prec t(c_2)$ or $t(c_2) \prec t(c_1)$. Assume $t(c_1) \prec t(c_2)$ (the proof in the second case is symmetric). We will actually prove that $c_1 \prec c_2$ follows in this case.

We thus have by our assumption and (1) of the lemma:

$$(7) \qquad\qquad q(c_1) \prec t(c_1) \prec t(c_2) \prec d(c_2) \prec c_2.$$

We are going to study now the two alternatives for $d(c_2)$ (derived from item (4) in the lemma), namely that $d(c_2)$ is a read of Q_1 and then that it is a read of $Turn$. In both cases we deduce that $c_1 \prec c_2$. At first, we give an informal outline of the proofs in each case.

In the first case, P_2 entered its critical section because it obtained $false$ in its last read of Q_1, and $d(c_2)$ is that read. The other process, P_1, wrote $true$ on Q_1 before c_1, and $q(c_1)$ denotes that write. Since $q(c_1) \prec d(c_2)$ there must be a write v of value $false$ by P_1 on Q_1 such that $q(c_1) \prec v \prec d(c_2)$. It follows that $q(c_1) \prec c_1 \prec v$, and hence that $c_1 \prec c_2$.

In the second case, P_2 entered its critical section because it obtained the value 1 in reading $Turn$. That is, $d(c_2) \prec c_2$ is a read of $Turn$. Yet P_2 itself wrote 2 on $Turn$ just before entering its 16–19 loop, and $t(c_2)$ denotes this write. So there is a write of 1 on $Turn$ (necessarily by P_1) denoted v such that $t(c_2) \prec v \prec d(c_2)$. Yet the write on $Turn$ preceding c_1, namely $t(c_1)$, satisfies $t(c_1) \prec t(c_2)$ and hence $t(c_1) \prec v \prec d(c_2) \prec c_2$. Yet $c_1 \prec v$ (because otherwise $v \prec c_1$, by seriality of P_1, implies that $t(c_1)$ is not the last write on $Turn$ preceding c_1). Hence $c_1 \prec c_2$.

We shall repeat the argument, but now checking that only properties listed in the lemma serve in the proof.

Case 1: $d(c_2)$ is a read of Q_1 of value $false$. That is $Read_{Q_1}(d(c_2)) \wedge Value(d(c_2)) = false$. Recall that ω is the return function defined on read events. So $w = \omega(d(c_2))$ is a write on Q_1 of value $false$. We are going to use $q(c_1) \prec d(c_2)$ derived from (7) above. As $q(c_1)$ is a write on Q_1, $q(c_1) \preceq \omega(d(c_2))$ follows. But $q(c_1)$ is a write of value $true$ and $d(c_2)$ is a read of value $false$ and hence

$$q(c_1) \prec \omega(d(c_2)).$$

It follows by (2) applied to $w = \omega(d(c_2))$ that $c_1 \prec w$, and as $\omega(d(c_2)) \prec d(c_2) \prec c_2$, we obtain $c_1 \prec c_2$ in this case.

Case 2: $d(c_2)$ is a read of *Turn* of value 1. We have $t(c_1) \prec t(c_2) \prec d(c_2)$ with $t(c_1)$, $t(c_2)$ being writes of 1 and 2 (respectively) on *Turn*, and $d(c_2)$ a read of *Turn* with $Value(d(c_2)) = 1$. Since $t(c_2)$ is a write on *Turn* and $d(c_2)$ a read, $t(c_2) \preceq \omega(d(c_2))$ follows from $t(c_2) \prec d(c_2)$, and even $t(c_2) \prec \omega(d(c_2))$ by the value difference. As $\omega(d(c_2))$ is a write of value 1 on *Turn*, it follows (from (3)) that $c_1 \prec \omega(d(c_2))$, and since $\omega(d(c_2)) \prec d(c_2) \prec c_2$ it follows again that $c_1 \prec c_2$ $\square$

The structure of this proof is typical of all correctness proofs given here. First a list of properties such as those of Lemma 1.1 is established and proved to hold in every history model. These properties speak only about the internal semantics of the processes. Then, in the second stage of the proof, the specifications of the communication devices are employed, and the correctness of the protocol is derived from the properties established in the first stage.

Two proofs of the mutual exclusion property were given: the assertional proof in Chapter 1 and this one, which is much longer, demands a certain background in model theory, and requires many concepts that were explained in three lengthy chapters. My reader certainly asks whether the price paid for rejecting the assertional method with its clear and simple interleaving model is not too high. I would like to suggest the following answer.

The proof in Chapter 1 is totally formal and I would be lost if asked to present its main lines informally. For the second protocol, I would define in words the functions q, t, and d, argue intuitively for the properties of the functions (stated in Lemma 1.1) and give the informal outline of the proof. I believe that most people when asked to give an informal argument for the mutual exclusion property will find something quite similar to the proof given here. Perhaps they will not use functions as I did and they may prefer to arrange the proof differently, but they will probably not reach first for the invariants of Chapter 1. In fact it took me several attempts over a couple of days to find these invariants, while it was matter of minutes to find an intuitive outline that developed into the present proof. It is true that this proof (in its detailed version) is longer and relies on more difficult concepts, but I do not think that any price is too high for keeping one's intuition. Having said that, I must also admit that different people may have different intuitions. It is not inconceivable that someone finds the first proof more to his intuitions, and there may well be some protocols whose proofs would look more natural even to me using global states. The investigation of concurrency requires good ideas, rather than a monolithic paradigm.

5
Higher-level events

*The protocol language of Chapter 3 is extended and procedure decla-
ration is included. This enables correctness proofs that argue about
higher-level events. First we give an example, the Pitcher/Catcher
Protocol of Chapter 3 (now in a procedural form), then a general
discussion, and finally two additional examples are analyzed: the
Kangaroo/Logaroo Protocol, and the Aimless Protocol.*

While computing machines execute instructions uniformly giving the same minute
attention to each of their thousand and one line codes, programmers need procedures and other higher-level constructs because they must *understand* their
programs. In its simplest form, a procedure call is just text substitution. A
procedure declaration associates the name of the procedure, R, with its body
S (which is a program with no procedure calls). Then whenever R appears in
a program, it is interpreted as a shorthand for S. We shall not deal here with
more complex forms (allowing parameter passing, local declarations, recursive
calls etc.) which make programming semantics such a difficult subject.

Even in this simplest form, procedure declaration allows programmers to organize programs and correctness proofs. The aim of this chapter is to show by
means of examples how to reason formally about higher-level procedures.

The assertional proof method, tied to the single-level analysis of changing
states, is closely related to the machine language level and not, so it seems, to the
higher-level events generated by the procedures. Lamport [20] emphasized that
system executions have the advantage of treating events at different granularities:
lower-level views are system executions that correspond to machine instruction
executions, and higher-level views are system executions with higher-level events
that correspond to procedure invocations. We adopt this approach, but want the
flexibility of having both higher-level and lower-level events in the same system
executions. This does not exclude the possibility of forming higher-level only or
lower-level only reducts.

procedure *PITCH*(a : N) begin (1) *obtain*(a); (2) *throw*(a) end	procedure *CATCH*(b, receive : N) begin (1) *catch*(b); (2) if *received* < b then *received* := b end

FIGURE 5.9. Declarations of *PITCH* and *CATCH*.

procedure *Pitcher*
{a is a natural number variable}
begin
repeat_forever
PITCH
end
	procedure *Catcher*
	{*received* and b are natural
	number variables}
	begin
	received := 0;
	repeat_forever
	CATCH
	end

protocol *Pitcher/ Catcher* concurrently do *Pitcher, Catcher* od	

FIGURE 5.10. Pitcher/Catcher with procedure calls.

1. *Pitcher/ Catcher* revisited

Recall the *Pitcher/Catcher* Protocol from Chapter 3, Section 4. We shall study a variant, emphasizing its procedural structure. The new version is in Figure 5.9, where two procedures *PITCH* and *CATCH* are declared, and in Figure 5.10 which gives the concurrent protocol in terms of these procedures, which clarifies the structure of the protocol.

Recall that our protocols have no variable declaration, but, instead, each language relies on a prefixed declaration of variables in each sort. Accordingly, when the background language L for the *Pitcher/ Catcher* protocol was described, variables a, *receive*, b were declared as natural number variables. Hence we view the statements that have the appearance of variable declaration as comments for the reader rather than part of the formal definition. Thus, for example, (a : N) following **procedure** *PITCH*, and the parenthetical {*received* and b are natural number variables} are comments. (The experienced reader surely can extend the semantics and allow variable declarations. We do not go in this direction, because our emphasis here is concurrency itself rather than programming, and so we view all variable declarations as informal remarks, made for the readers' convenience.)

The value of variable a is determined by some external operation *obtain* which assigns a natural number to a. *obtain* is always terminating, but other than that no restriction is imposed on its possible values.

Let us pause for a moment and indicate what we want to prove. A *CATCH* operation for which condition "*receive* < b" holds is said to be *successful*. Such

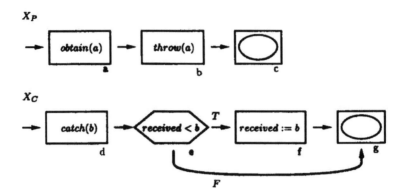

FIGURE 5.11. Flowcharts X_P and X_C of the *PITCH* and the *CATCH* procedures.

operations increase the value of *receive* up to b, while unsuccessful operations do not change it. Now to each successful *CATCH* operation C we define the corresponding *PITCH* operation, denoted $\Phi(C)$. Informally, $\Phi(C)$ is the *PITCH* operation containing the "*throw(a)*" execution which "*catch(b)*" obtained in C. Our aim is to prove that Φ, as a function from the successful higher-level *CATCH* events to the *PITCH* events, is monotonic (in the precedence ordering) if the values thrown by *Pitcher* never decrease. The informal argument is quite easy. Assume that $C_1 \prec C_2$ are two successful *CATCH* events, and v_1, v_2 denote the values of the *catch* events in C_1 and C_2 respectively. Since $C_1 \prec C_2$, $v_1 < v_2$ (because C_1 sets the value of *receive* to v_1, and C_2 obtained an even higher *catch*). If $P_1 = \Phi(C_1)$ and $P_2 = \Phi(C_2)$ are the corresponding *PITCH* events, then v_1 and v_2 are the values thrown by P_1 and P_2 (respectively), and as $v_1 < v_2$, $P_1 \prec P_2$ follows since otherwise $P_2 \prec P_1$ would imply $v_2 \leq v_1$ by our assumption (that the values thrown by *Pitcher* never decrease).

The formal, mathematical proof that we shall give for the monotonicity of Φ follows this outline (Lemma 1.6), but is necessarily longer because it relies on the exact definition of the semantics of the protocol. It will take us a while before we can formally define higher-level *PITCH/CATCH* events. We must first define the flowcharts for the protocol, consider its histories and system executions (incorporating the *throw/catch* specifications) and only then can the proof be followed. Part of that proof deals with the higher-level *PITCH/CATCH* events, and that is our reason for bringing this example here—to illustrate how a formal higher-level argument can be conducted.

PITCH and *CATCH* are serial procedures and their flowcharts, given in Figure 5.11, follow the definition of Chapter 3. The flowchart of *PITCH* is called X_P, and that of *CATCH* is called X_C.

The higher-level flowcharts H_P and H_C of the *Pitcher* and the *Catcher* procedures are given in figure 5.12. Observe that nodes 2 and 6 are labeled with the names of the invoked procedures and their variables, which indicates that we treat these procedure names as external operations.

We define now *states*(H_P), the set of states of the higher-level *Pitcher* flowchart

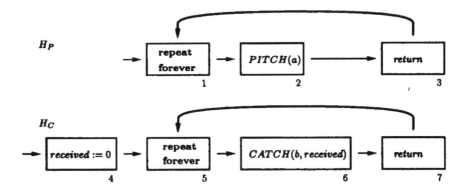

FIGURE 5.12. Higher-levels flowcharts of the *Pitcher* and *Catcher* procedures.

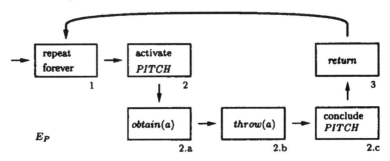

FIGURE 5.13. Flowchart E_P is an explicit expansion of H_P.

H_P. $\mathcal{V}(H_P) = \{a\}$. Let C be the location variable. A regular state over H_P is a function S defined on $\{C, a\}$ with $S(C) \in \{1, 2, 3\} = locations(H_P)$ and $S(a) \in \mathbf{N}$. An unreachable state U is a function that is only defined on C and is such that $U(C) \in locations(H_P)$.

S is an initial state of H_P if $S(C) = 1$ and $S(a)$ is any value.

Transitions are defined as in Chapter 3. In particular to arrow $\alpha = (2, 3)$, the $PITCH(a)$ arrow, we define $(S_1, S_2) \in \mathcal{T}r(\alpha)$ iff S_1, S_2 are states with $S_1(C) = 2$ and $S_2(C) = 3$. So S_2 can be unreachable, and then (S_1, S_2) is a non-terminating transition, or else S_2 is a regular state and $S_2(a)$ is any natural number.

In this semantics, $PITCH(a)$ is seen as some undefined external operation called with a variable parameter a. Yet we know that $PITCH(a)$ is implemented by the $PITCH$ flowchart X_P. To incorporate this additional information on how $PITCH(a)$ is implemented, we consider the explicit expansion of the pitcher flowchart, called E_P. This explicit flowchart is obtained, roughly speaking, by grafting the $PITCH$ flowchart, X_P, into the $PITCH(a)$ arrow (the $(2, 3)$ arrow) of H_P. The result is illustrated in Figure 5.13.

We now discuss histories. A history of the explicit flowchart E_P is a finite or infinite sequence σ such that:

(1) For every i, $S = \sigma(i)$ is an E_P state. That is, S is either a regular state and $S(C) \in \{1, 2, 2.\text{a}, 2.\text{b}, 2.\text{c}, 3\} = locations(E_P)$ and $S(a) \in \mathbb{N}$, or else S is an unreachable state assigning to C a node in $locations(E_P)$ (and then i is the last index of σ). In fact, the only possible location for the unreachable state is 2.c, since $throw(a)$ is the only possibly non-terminating operation in E_P. (We have assumed that $obtain$ is always terminating.)

(2) The first state $S_1 = \sigma(1)$ is an initial state (i.e., $S_1(C) = 1$). (We start the sequence with $i = 1$ because $\sigma(0)$ is reserved for the global initial state of the combined flowchart.)

(3) For every i such that $\sigma(i)$ and $\sigma(i+1)$ are defined, $\langle \sigma(i), \sigma(i+1) \rangle$ is an E_P transition.

Item (3) needs some further explanation because the flowchart contains two new labels: "activate $PITCH$" and "conclude $PITCH$" whose transitions have never been defined. Arrow $(2, 2.\text{a})$ is called an *activation arrow* and it represents the call to procedure $PITCH$. Arrow $(2.\text{c}, 3)$ is called a *conclusion arrow* and it represents the resumption of the calling procedure ($Pitcher$). Formally, the activation transitions $Tr(2, 2.\text{a})$ are all regular transitions that change the value of C from 2 to 2.a (but do not change the value of variable a). The conclusion transitions $Tr(2.\text{c}, 3)$ are similarly defined.

A history σ for E_P can be finite. This happens exactly when for some i $\sigma(i) = u$ is an unreachable state. Necessarily $u(C) = 2.\text{c}$ in this case, and i is the maximal index of σ.

Any history of E_P induces a history of H_P. To define this induced history we first discuss the relationship between states of E_P (the explicit states) and states of H_P (higher level states). Some of the states in $states(E_P)$ can be viewed as states of the higher-level flowchart H_P. Since $\mathcal{V}(E_P) = \mathcal{V}(H_P) = \{a\}$, if $S \in states(E_P)$ is (a regular state) such that $S(C) \in \{1, 2, 3\}$, then we can view S as a H_P state (by identifying nodes $1, 2, 3$ with those nodes in H_P that have the same number name). We say that $S \in states(E_P)$ *induces* the corresponding H_P state (which we continue to denote with S). The unreachable E_P state u with $u(C) = 2.\text{c}$ (representing the outcome of a non-terminating $throw(a)$ operation) also induces a H_P state, namely the unreachable H_P state U with $U(C) = 3$ (representing the outcome of a non-terminating $PITCH$ call).

Now, if σ is an E_P-history then the induced H_P-history $\sigma \mid H_P$ is defined by picking from σ only those states $\sigma(i)$ that induce H_P-states and by forming the sequence of induced states. We shall formally define $\tau = \sigma \mid H_P$ by defining inductively an increasing sequence of indices $\langle n_i \mid i \in \omega \rangle$ (or $\langle n_i \mid i < i_0 \rangle$ if σ is finite) and defining $\tau(i)$ to be the H_P state induced by $\sigma(n_i)$. Inductively, we shall have $\sigma(n_i)(C) \in \{1, 2, 3\}$. The induction goes as follows.

(1) $n_1 = 1$ and $\tau(1) = \sigma(1)$ is the initial state of σ.

(2) Suppose that $\tau(i) = \sigma(n_i)$ is defined. Since $\sigma(n_i)(C) \in \{1, 2, 3\}$, the following possibilities exist.

(a) $\sigma(n_i)(C) \in \{1, 3\}$. Then $n_{i+1} = n_i + 1$, and $\tau(i+1) = \sigma(n_{i+1})$.

(b) $\sigma(n_i)(C) = 2$. There are two cases here.

A: $\sigma(n_i + 3)$ is a regular state. That is, the transition

$$\langle \sigma(n_i + 2), \sigma(n_i + 3) \rangle$$

corresponding to the *throw(a)* instruction is terminating. Then we define $n_{i+1} = n_i + 4$ and so $\tau(i + 1) = \sigma(n_i + 4)$. That is, $\tau(i + 1)$ is the first state reaching back the H_P flowchart (at node 3).

B: $\sigma(n_i + 3)$ is an unreachable state. Then $\tau(i + 1)$ is the unreachable H_P state U with $U(C) = 3$. (The sequence τ ends then with $\tau(i + 1)$.)

In all cases $\langle \tau(i), \tau(i + 1) \rangle$ is an H_P transition. So τ is an H_P history. We say that $\tau = \sigma \mid H_P$ is the higher-level history induced by σ.

For example, suppose that σ as in Table 1 contains 11 states. These induce 6 states in the H_P history τ. Roughly speaking the σ states related to the X_P nodes 2.a, 2.b, 2.c are omitted, and the remaining six states form τ. This description is somewhat inaccurate for three reasons:

(1) The states of E_P are not states of H_P. Only if nodes 1, 2, 3 of E_P are identified as nodes of H_P then such a translation can be done.

(2) Omitting members from a sequence does not lead to a sequence. We must define n_i if we want τ to be a sequence(defined on successive natural numbers). In our case $n_1 = 1$, $n_2 = 2$, $n_3 = 6$ etc.

(3) If the last state in σ is an unreachable state (as it is in our example), then it is not omitted, but is rather replaced with the H_P unreachable state located at node 3.

Recall (from Chapter 3) that steps are representations of transitions. If σ is an E_P history, then the ith step in σ represents the transition $\langle \sigma(i), \sigma(i + 1) \rangle$ (when $i + 1$ is in the domain of σ). What is the relationship between the steps of σ and those of the induced history $\tau = \sigma | H_P$? The substep relation is defined naturally as follows. If $n_i + 1 < n_{i+1}$, then the kth step of σ for $n_i \leq k < n_{i+1}$ is a substep of the ith step of τ. And if σ is finite, then the last step, the non-terminating "*throw(a)*" step, is a substep of the non-terminating *PITCH* step. The meaning should be clear, even if it takes a minute or two to clarify the indices. For example, referring to Table 1, steps $\langle P, 2 \rangle$, $\langle P, 3 \rangle$, $\langle P, 4 \rangle$, $\langle P, 5 \rangle$ of σ are all substeps of the second H_P step (denoted $\langle H_P, 2 \rangle$).

Perhaps it is illuminating to consider the transitions rather than the steps. Well, all transitions in between an activation transition and its pairing conclusion transition (including these activation/conclusion transitions) are considered to be sub-transitions of the corresponding higher-level τ transition. In addition, if σ contains a non-terminating transition, then this transition (and the preceding transitions after activation) are sub-transitions of the corresponding non-terminating τ-transition.

For example, in table 1, σ consists of 11 states, and thus 10 steps are defined. $\sigma(1)$ is the first state, its location (under column C) is node 1, and its value at variable a is (well I had to choose a number) 7. The first step is denoted $\langle P, 1 \rangle$ and it represents transition $\langle \sigma(1), \sigma(2) \rangle$ which is called "repeat forever" as this is the label on node 1. Still in the first line we read that state $\tau(1)$ is induced by $\sigma(1)$. There are six induced states, and five higher-level steps. If we use $\in$ to

state	C	a	step	transition	H_P state	step	transition
$\sigma(1)$	1	7	$\langle P,1 \rangle$	repeat forever	$\tau(1)$	$\langle H_P,1 \rangle$	repeat forever
$\sigma(2)$	2	7	$\langle P,2 \rangle$	activation	$\tau(2)$	$\langle H_P,2 \rangle$	PITCH
$\sigma(3)$	2.a	7	$\langle P,3 \rangle$	obtain(a)			
$\sigma(4)$	2.b	1	$\langle P,4 \rangle$	throw(a)			
$\sigma(5)$	2.c	1	$\langle P,5 \rangle$	conclusion			
$\sigma(6)$	3	1	$\langle P,6 \rangle$	return	$\tau(3)$	$\langle H_P,3 \rangle$	return
$\sigma(7)$	1	1	$\langle P,7 \rangle$	repeat forever	$\tau(4)$	$\langle H_P,4 \rangle$	repeat
$\sigma(8)$	2	1	$\langle P,8 \rangle$	activation	$\tau(5)$	$\langle H_P,5 \rangle$	PITCH
$\sigma(9)$	2.a	1	$\langle P,9 \rangle$	obtain			
$\sigma(10)$	2.b	3	$\langle P,10 \rangle$	throw(a)			
$\sigma(11)$	2.c		unreachable state		$\tau(6)$		Unreachable state 3

TABLE 5.1. The state table of the explicit history σ and the induced higher-level history τ $\tau = \sigma \mid H_P$. The initial value of a is assumed to be 7. Step $\langle P,10 \rangle$ is a non-terminating $throw(a)$ step. $\langle H_P,5 \rangle$ is a non-terminating $PITCH$ step

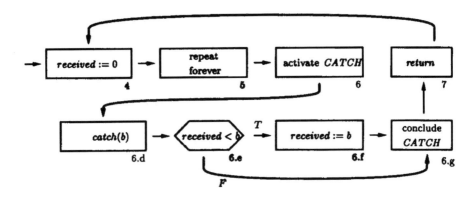

FIGURE 5.14. Explicit expansion of H_C into E_C.

denote the substep relation, then

$$\langle P,2 \rangle, \langle P,3 \rangle, \langle P,4 \rangle, \langle P,5 \rangle \in \langle H_P,2 \rangle,$$

$$\langle P,8 \rangle, \langle P,9 \rangle, \langle P,10 \rangle \in \langle H_P,5 \rangle.$$

We discuss the corresponding development, but now with less detail, for the *Catcher* protocols and the combined protocols. Flowcharts X_C and H_C were drawn in figures 5.11 and 5.12, and the explicit flowchart E_C is in figure 5.14

As before, some of the E_C states induce H_C states. These are the regular E_C-states S with $S(C) \in \{4,5,6,7\}$ which can be seen as H_C states, and the unreachable E_C-state u with $u(C) = 6.e$, inducing the unreachable H_C-state U with $U(C) = 7$ ($catch(b)$ may be non-terminating). Again, if σ is an E_C history, then $\sigma \mid H_C$ is the induced H_C history.

Now we form the combined flowcharts: the combined higher-level H_P/H_C flowchart (Figure 5.15), and the combined explicit E_P/E_C flowchart (Figure

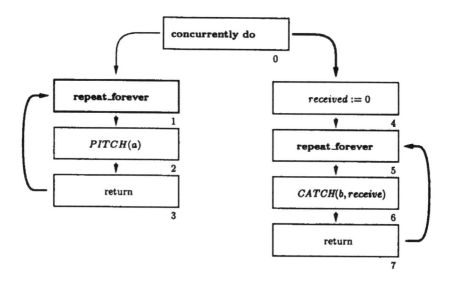

FIGURE 5.15. The combined higher-level H_P/H_C flowchart.

5.16).

Let $H = (\sigma_P, \sigma_C)$ be a history for the combined explicit E_P/E_C flowchart. That is, σ_P and σ_C are sequences of states, σ_P is an E_P history and σ_C is an E_C history, such that $\sigma_P(0) = \sigma_C(0)$ is the same global initial state. In fact, the history (as defined in Chapter 3) and its flattened history model are richer structures in which steps and transitions are definable. Recall that the steps are pairs: $\langle P, i \rangle$ is a *Pitcher* step, and $\langle C, j \rangle$ is a *Catcher* step. $\langle P, i \rangle$ represents the transition $(\sigma_P(i), \sigma_P(i+1))$ of *Pitcher*, and $\langle C, j \rangle$ represents the transition $(\sigma_C(j), \sigma_C(j+1))$.

We further expand the history and represent the induced higher-level histories $\tau_P = \sigma_P|H_P$, $\tau_C = \sigma_C|H_C$. Then (τ_P, τ_C) is a H_P/H_C history which we denote by $H|H_P/H_C$. In the two-level history model of $(\sigma_P, \sigma_C, \tau_P, \tau_C)$ we can define the *substep relation* between σ_P steps and τ_P steps (and between σ_C steps and τ_C steps). This allows us to collect sets of steps that represent executions of the *PITCH/CATCH* procedures. Namely, for each *PITCH* step $\langle H_P, i \rangle$ in τ_P (representing some $(2,3)$ transition) we collect all the σ_P substeps into a set X, and we say that X forms an execution of the *PITCH* procedure. We write *PITCH_event*(X) in this case. In a similar fashion the predicate *CATCH_event*(X) is defined, and the set of σ_C steps X is said to form an execution of the *CATCH* procedure.

Recall the ordering of the steps:

$$\langle P, i \rangle <_{steps} \langle P, i+1 \rangle$$

$$\langle H_P, i \rangle <_{steps} \langle H_P, i+1 \rangle.$$

The steps of *Catcher* are similarly ordered. Then $<_{steps}$ is extended on sets of

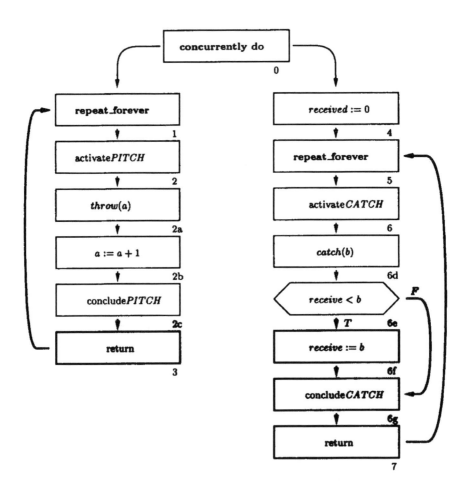

FIGURE 5.16. The combined explicit E_P/E_C flowchart.

steps:

$$X_1 <_{steps} X_2 \text{ iff } \forall s \in X_1 \ t \in X_2 (s <_{steps} t).$$

If $PITCH\_event(X)$ (and similarly if $CATCH\_event(X)$) then X is clearly a convex interval of *Pitcher*'s steps. In fact, as H_P is a serial flowchart, the *PITCH* steps in τ_P are linearly ordered, and if $\langle H_P, i \rangle, \langle H_P, j \rangle$ are *PITCH* steps, and X_0 X_1 are the corresponding executions then $\langle H_P, i \rangle <_{steps} \langle H_P, j \rangle$ iff $X_0 <_{steps} X_1$ iff $i < j$.

DEFINITION 1.1. *Let* $\sigma_P, \sigma_C, \tau_P, \tau_C$ *be as above, and let* H *be the corresponding history model. Suppose that* $CATCH\_event(X)$ *holds in* H.

(1) *Let* $rec(X)$ *be the value of variable received at the entry to* X, *that is, the value before executing line (2) in the* CATCH *procedure (of Figure 5.9). And (in case* X *is terminating) let* $rec'(X)$ *be the value of received after line (2) is executed in* X. *It is possible that* $rec(X) = rec'(X)$ *and that* $rec(X) < rec'(X)$, *the latter case occuring exactly if the condition received* $< b$ *holds in* X.

 A more formal definition of $rec(X)$ *should use the definition of* X *as a set of* E_C *steps. Suppose* $\langle C, i \rangle$ *is the first step in* X, *the activation step, and (if* X *is terminating)* $langle C, j \rangle$ *is the last step in* X. *Then we define* $rec(X) = \sigma_C(i)(received)$, *and if* X *is terminating, then we also define* $rec'(X) = \sigma_C(j)(received)$.

(2) *We say that* X *is successful, and write* $Successful(X)$, *if the condition "received* $< b$*" holds in* X *in executing line (2). Then* $Successful(X)$ *holds iff (* X *is terminating and)* $rec(X) < rec'(X)$.

(3) *If* X *in* S *is a terminating* CATCH *or a terminating* PITCH *event, then* $Val(X)$ *denotes the value of the catch or respectively the throw event in* X. *(Equivalently, the values of variables* b *or* a, *after executing line (1).)*

It can easily be seen that if $CATCH\_event(X) \wedge Terminating(X)$ then $Val(X) \leq rec'(X)$, and if also $Successful(X)$ holds, then

(8) $$rec(X) < Val(X) = rec'(X)$$

holds.

Let $H = (\sigma_P, \sigma_C)$ be an E_P/E_C history as above, and let $\mathcal{H}$ be a history model flattening of H. The induced histories τ_P and τ_C can be defined in $\mathcal{H}$, and the substep relation is used to define $PITCH/CATCH$ executions (falling under predicates $PITCH\_event/CATCH\_event$).

Let $C_0, \ldots, C_i \ldots$ enumerate all the $CATCH$ events in S, in $<_{steps}$ increasing order. So either this sequence is infinite, or else it is defined only for $i = 0, \ldots, i_0$ (and then C_{i_0} is nonterminating).

Since the $CATCH$ executions and the $PITCH$ executions are serially ordered, we may speak about successors: X_2 is the successor of X_1 if both are, say, $CATCH$ executions, $X_1 <_{steps} X_2$, and there is no $CATCH$ execution X in between. So, if $CATCH\_event(X_1)$ and there is any $CATCH$ event at all after X_1, then there is also a successor to X_1.

Now suppose that X_1, X_2 are $CATCH$ events and X_2 is the successor of X_1. Then $rec'(X_1) = rec(X_2)$. That is, the final value of received at X_1 becomes the initial value at the successor event. Since only successful $CATCH$ events change the value of $received$, we have the following.

LEMMA 1.2.   (1) If $X = C_i$ is terminating then

$$rec'(X) = \max\{rec(X), Val(X)\}.$$

(2) If C_i and C_{i+1} are defined, then $rec'(C_i) = rec(C_{i+1})$, and so

$$rec(C_i) \leq rec(C_{i+1})$$

follows from (1). Moreover, if C_i is successful then

$$rec(C_i) < Val(C_i) = rec(C_{i+1}).$$

(3) If C_i and C_j are defined and $i < j$, then (1) and (2) imply

$$Val(C_i) \leq rec'(C_i) \leq rec(C_j),$$

so if in addition C_j is successful then $Val(C_i) < Val(C_j)$ follows from $rec(C_j) < Val(C_j)$.

Let S be a system execution extension of $\mathcal{H}$. Thus (as in Chapter 3 Definition 2.2) the steps of σ_P and σ_C are the events, and the precedence relation $\prec$ extends the $<_{steps}$ ordering. For emphasis, and in view of the higher-level events to be introduced soon, we call these events "lower-level". The induced histories τ_P and τ_C are defined in S, and the substep relation is used to define higher-level events in S, that is sets of lower-level events. Namely, $PITCH$ and $CATCH$ higher-level events are formed as in page 1. They are also called $PITCH$, $CATCH$ operation executions.

Extend the precedence relation $\prec$ of S on these higher-level sets as in Chapter 2 Section 3. Namely

$$X \prec Y \text{ iff } \forall x \in X \forall y \in Y (x \prec y).$$

$\mathcal{H}$ is thus transformed into a two-level system execution, containing lower-level and higher-level events related with the membership relation $\in$. We further assume that the *throw/catch* events satisfy the specifications of chapter 2 (page 40), expressed with the function α.

The main definition can now be made.

DEFINITION 1.3. *If $CATCH\_event(X) \wedge Terminating(X)$ holds, then define $\Phi(X)$ as the following PITCH event: Let $c \in X$ be the terminating catch event, look at $\alpha(c)$ (the corresponding throw event) and find PITCH event Y such that $\alpha(c) \in Y$. Then define*

$$Y = \Phi(X).$$

The fact that every throw event in $\mathcal{H}$ belongs to some uniquely determined PITCH operation execution is used in this definition naturally.

We may now state our aim: To prove that Φ is monotonic on the successful $CATCH$ events, if $Val(X)$ is weakly monotonic on the $PITCH$ events. That is, to show that the following holds in S:

(1) The *PITCH* events are serially ordered, and the *CATCH* events are serially ordered.

(2) If C is a terminating *CATCH* event, then $P = \Phi(C)$ is a terminating *PITCH* event and $Val(\Phi(C)) = Val(C)$.

(3) If $C_1 \prec C_2$ are *CATCH* events and C_2 is successful, then $Val(C_1) < Val(C_2)$.

FIGURE 5.17. Higher-level properties.

THEOREM 1.4. *Suppose that for every two PITCH events, $P_1 \prec P_2$ implies $Val(P_1) \leq Val(P_2)$. If $X_1 \prec X_2$ are two successful CATCH events, then $\Phi(X_1) \prec \Phi(X_2)$.*

The proof is in the following two lemmas.

LEMMA 1.5. *For any terminating CATCH event X,*

$$Val(\Phi(X)) = Val(X).$$

Proof. The proof follows the definition of Φ and *Val*, and the assumption that $Value(c) = Value(\alpha(c))$ for every terminating *catch* event c. Indeed, let X be a terminating *CATCH* event. Let $c \in X$ be the *catch* event in X. Then c is terminating. By definition, $Y = \Phi(X)$ is the *PITCH* event containing $\alpha(c)$. The definition of *Val* is such that $Val(X) = Value(c)$, and $Val(Y) = Value(\alpha(c))$. As $Value(c) = Value(\alpha(c))$, the required equality follows. $\square$

Figure 5.17 now summarizes all higher-levels properties so far established. These are statements in a first-order language containing predicates *PITCH_event*, *CATCH_event*, and *Successful*; functions *Val*, and Φ, and the ordering $\prec$. (For clarity, we say:

> for every *PITCH* event...

rather than

> for every X if $PITCH\_event(X)$ then....

LEMMA 1.6. *Assume the higher-level properties of Figure 5.17 (and the obvious axioms concerning the ordering of the natural numbers). Suppose that for every two terminating PITCH events P_1 and P_2, $P_1 \prec P_2$ implies $Val(P_1) \leq Val(P_2)$. Then Φ is monotonic on the successful CATCH events: If $C_1 \prec C_2$ and $Successful(C_2)$, then $\Phi(C_1) \prec \Phi(C_2)$.*

Proof. Suppose that $C_1 \prec C_2$ and C_2 is successful. Then $Val(C_1) < Val(C_2)$ by (3). Say $P_1 = \Phi(C_1)$, and $P_2 = \Phi(C_2)$. Then $Val(P_i) = Val(C_i)$ for $i = 1, 2$. Hence $Val(P_1) < Val(P_2)$. But this implies $P_1 \prec P_2$ since otherwise $P_2 \preceq P_1$ (by seriality) would imply $Val(P_2) \leq Val(P_1)$ by our assumption. Hence $P_1 \prec P_2$. $\square$

I admit that this was quite a long proof for such a simple protocol, but it served its purpose to illustrate a method and to introduce new concepts. In the future some of the formal steps (such as defining the higher-level events) will be done more casually, relying on our intuitions. The reader can then supply the missing details and obtain a completely formal proof.

The proof given here is typical in that it is divided into two phases. In the first, the higher-level properties (such as those listed in figure 5.17) are obtained, and in the second, consequences of these properties are derived. In the second phase the protocols are completely forgotten, and only the abstract properties obtained in the first stage are used.

In my experience, this division not only allows a better understanding of the protocol, but it also leads to better protocols. It is often the case that the proofs of the second phase suggest the possibility of better higher-level properties. Then the programmer returns to the protocols and tries to change them accordingly. This tuning may take several rounds, and results in improved protocols.

2. Procedure calls

We summarize here in general terms our treatment of procedure calls, as exemplified in the preceding section. Our protocol language is extended in order to allow for procedure calls. A procedure declaration has the form

procedure proc_name **begin** proc_body **end**

where proc_name is the name of the procedure, and proc_body is a serial instruction as defined in Chapter 3 (and in particular no procedure calls are possible in this instruction).

If $D_1, \ldots, D_k$ are the procedure names of k procedures declarations, and if the list of variables appearing in the body of D_i is $x_{i_1}, \ldots, x_{i_\ell}$ (where ℓ may depend on i) then an "extended serial instruction" is defined with the addition of external instructions of the form

$$D_i(x_{i_1}, \ldots, x_{i_\ell}).$$

That is, we allow procedure calls only with the original list of variables. Now (concurrent) protocols are defined as before with the following form.

protocol name **concurrently** do list of extended serial instructions **od**

An example of a concurrent protocol was given in Figure 5.10

protocol *Pitcher/Catcher*
concurrently do *Pitcher, Catcher* od

The list of serial instructions here consists of only two instructions, namely *Pitcher*, and *Catcher* which call procedures *PITCH* and *CATCH*.

A concurrent protocol has two flowcharts now:

(1) The combined higher-level flowchart H in which external instructions $D(x_1, \ldots, x_\ell)$ appear as tags. (For example Figure 5.15.)

(2) The explicit form E (for example Figure 5.16) in which any

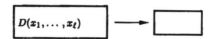

is replaced by

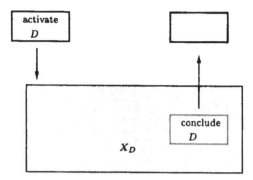

where X_D is the serial flowchart of procedure D with its final node marked "conclude D".

Some of the nodes of the explicit flowchart E are in fact nodes of the higher-level flowchart H, while other nodes come from the flowcharts X_D of the "grafted" procedures. Correspondingly, any history σ of E induces a history $\tau = \sigma|H$ of H. Visually, σ traces a path in E (namely the sequence of nodes $\langle \sigma(i)(C) \mid i \in \mathbb{N} \rangle$) and when the nodes of H are marked-out in this sequence, the resulting H-path induces an H history. The precise mathematical definition is left to the reader.

What is the aim of this explicit and induced history? To enable the definition of higher-level events that represent procedure executions. The collection of all substeps (in E) of a D step (in H) forms an execution of the D procedure. This execution can be terminating or not. It is terminating exactly if it is a finite set of terminating steps.

If S is a system execution of flowchart E (namely S extends history σ), then the precedence relation $\prec$ can be naturally extended on the higher-level events thus defined by

$$X \prec Y \text{ iff } \forall x \in X \forall y \in Y (x \prec y)$$

for any higher-level events X and Y, and by

$$X \prec a \text{ iff } \forall x \in X (a \prec x),$$

$$a \prec X \text{ iff } \forall x \in X (x \prec a).$$

Now we can argue about the higher-level events in S (which is said to be a two-level system-execution).

3. KanGaroo and LoGaroo

In this section another example is given for the use of higher-level events to reason about the correctness of protocols. We shall be much less formal in the first presentation, and indicate the missing details in the following subsection.

The example (see Figure 5.18) assumes $p_0 + 1$ processes,

$$KanGaroo_1, \ldots, KanGaroo_{p_0}$$

(collectively called $KanGaroo$), and $LoGaroo$, sharing an array $A[i]$, $0 \le i < N$, of boolean serial registers. The set of indices $[0, \ldots, N-1]$ is called $TagNumber$. All processes can read and write on these registers, and it is assumed that the hardware somehow ensures mutual exclusion. That is, we assume seriality of each

procedure *Kan*	procedure *Lo*
var t : *TagNumber*;	var *frame_expected*: Natural number;
begin	x : boolean;
	begin
(1) obtain_a_tag(t);	
(2) *Write*(A[t], true)	(1) repeat
end	$Read(A[[frame\_expected]_N], x)$
	until x = true;
	(2) $Write(A[[frame\_expected]_N], false)$;
	(3) *frame_expected* := *frame_expected* + 1
	end
procedure *KanGaroo*	procedure *LoGaroo*
begin	begin
repeat_forever	*frame_expected* := 0;
Kan	repeat_forever
end	*Lo*
	end
protocol *KanGaroo/LoGaroo*	
concurrently do *KanGaroo, LoGaroo* od	

FIGURE 5.18. *KanGaroo* and *LoGaroo*. A[i] are initially **false**.

A[i]. *Write*(A[i], v) is the instruction to write value v on A[i], and *Read*(A[i], x) is the reading instruction, which determines the value of variable x.

KanGaroos and *LoGaroo* operate concurrently. Each *KanGaroo* executes forever one after the other *Kan* operation, which consist in:

(1) Obtaining some tag number t. Thus obtain_a_tag is an external, not necessarily terminating, operation which somehow—we do not care how—determines a value assigned to variable t.
(2) Writing **true** on register A[t].

KanGaroo thus never reads the registers and only writes **true** on some arbitrarily determined A[t] (possibly writing many times on the same register).

LoGaroo, on the other hand, writes only **false**, but only after reading **true**. *LoGaroo* is very organized, he has a variable *frame_expected* which is initialized to zero, and is increased by one in each terminating *Lo* operation. First, *Lo* waits until $A[[frame\_expected]_N]$ is *true*, and then it resets it to **false**, and finally it increases *frame_expected*. (For any number j, $[j]_N = j \pmod N$ is the tag number congruent to j mode N.)

Let S be a system execution of the *KanGaroo/LoGaroo* Protocol in which registers A[i] are all serial.

As before (Chapter 2 section 4), we assume a function ω defined on the read events and giving for each read event r of A[i], a write event $\omega(r)$ which is the $\prec$-rightmost write on A[i] preceding r. Of course, $Value(\omega(r)) = Value(r)$ always holds. Assume an initial value **false** for each of these registers.

An execution of procedure *Lo* is said to be *successful* just in case the **repeat** loop is terminating, that is there is a read of register $A[[frame\_expected]_N]$ that returned *true*. This read is called the *successful* read of the execution. A successful execution is necessarily terminating, since all other instructions in *Lo* are terminating.

Recall from Section 2 that a call to *Kan* (*Lo*) is modeled by the set of all sub-steps pertaining to that call, namely all lower-level events between the activation and the following conclusion events. Clearly some of these events are more important than the others, more relevant to the correctness proof, and I want to form higher-level events that contain only these relevant events. For this I will define higher-level *Kan* (*Lo*) *events* (rather than executions). The terminology event/execution is somewhat arbitrary, but the distinction is useful. Specifically, the higher-level events are defined in S as follows.

(1) A non-terminating *Kan* event contains a single event o which is the non-terminating execution of the obtain_a_tag instruction. A terminating *Kan* event consists of two lower-level events: $X = \{o, w\}$ is a *Kan* event if o is the terminating obtain_a_tag event and w is the write event, corresponding to lines (1) and (2) in some terminating *Kan* execution. If $0 \le t < N$ is the tag number obtained by o then we define

$$t = tag(X).$$

In this case w is a write on register $A[t]$.

(2) A *Lo* event corresponds to an execution of the *Lo* procedure as follows.

  (a) If the execution is not successful, then it contains infinitely many read events, all returning *false*, of register $(A[\lceil frame\_expected\rceil_N]$. The corresponding higher-level event X is this infinite set of read events, and we write $Lo(X) \wedge \neg Terminating(X)$.

  (b) If the execution is successful, then it contains a successful read r of register $A[\lceil frame\_expected\rceil_N]$ followed by a write w (of value *false*) on that same register. We form $X = \{r, w\}$ and define $Lo(X)$. Thus, even if the execution of the successful *Lo* operation contains several reads that returned the value *false*, the corresponding higher-level event X contains only the successful read (and the following write).

(3) In addition to these *Kan*/*Lo* events, initializing events are also defined. these are the writes of *false* on registers $A[i]$.

(4) We make a definition: If $Lo(X)$, then $frame\_expected(X)$ denotes the value of variable $frame\_expected$ at the entry to X.

Since the higher-level *Lo* events are serially ordered in S, we can enumerate them in increasing order $\langle L_j \mid j \in J_0 \rangle$ where $J_0 \subseteq$ N, the index set, is either N (the set of natural numbers) or an initial segment of N. L_0 is the first *Lo* event, and $L_j \prec L_{j+1}$ whenever $j + 1 \in J_0$. If J_0 is infinite, then all L_j are terminating (equivalently, successful), but otherwise, if j_0 is the maximal index in J_0, then J_{j_0} is the last *Lo* event. J_{j_0} is not successful.

Clearly, for every $j \in J_0$, $frame\_expected(L_j) = j$. Indeed $frame\_expected(L_0) = 0$ because 0 is the initial value of $frame\_expected$. Since $frame\_expected$ is increased by 1 at line (3), each successful *Lo* event increases $frame\_expected$ by 1. Hence $frame\_expected(L_{i+1}) = frame\_expected(L_i) + 1$. Induction gives

$$frame\_expected(L_i) = i.$$

THEOREM 3.1. *Suppose that J_0 is finite, and $j_0 = \max(J_0)$. Then L_{j_0} is non terminating. If for some terminating Kan event, K, $tag(K) = \lceil j_0 \rceil_N$, then $j_0 \geq N$ and it is not the case that $L_{j_0 - N} \prec K$.*

Proof. We know that $frame\_expected(L_{j_0}) = j_0$. Since L_{j_0} is not terminating, it consists of an infinite number of read events of $A[\lceil j_0 \rceil_N]$, all returning *false*. On the other hand $K = \{o, w\}$ is terminating, and by assumption

$$t = tag(K) = j_0 \quad (\text{mod } N).$$

Thus w is a write of value *true* on $A[t]$. By the finiteness property of Lamport, there is a read $r \in L_{j_0}$ such that $w \prec r$. Since w is a write of *true* in K, and r a read returning false,

$$w \prec \omega(r)$$

necessarily holds (because the alternative possibilities $w = \omega(r)$ and $\omega(r) \prec w$ are impossible). So $\omega(r)$ is a write of value *false* on $A[t]$ and it is not the initial write. Hence $\omega(r)$ belongs to the rightmost Lo event $L_j \prec L_{j_0}$ which writes (*false*) on $A[t]$, and so there is $0 \leq j < j_0$ with $j = j_0$ (mod N). This implies $j_0 \geq N$. Since $L_{j_0 - N}$ is the rightmost $L_j \prec L_{j_0}$ with $j = j_0$ (mod N), $\omega(r)$ is in $L_{j_0 - N}$. Hence $L_{j_0 - N} \prec K$ is impossible, as $w \in K$ and $\omega(r) \in L_{j_0 - N}$ would imply $\omega(r) \prec w \prec r$. $\square$

For the second theorem that is proved here we need the following definition of $\Gamma : Lo \to Kan$ defined on the successful Lo events. Let L_j be a successful Lo event, and let $r \in L_j$ be its successful read of $A[\lceil frame\_expected \rceil_N]$. Since $frame\_expected(L_j) = j$, r is a read of $A[\lceil j \rceil_N]$. Let $w = \omega(r)$ be the corresponding write on $A[\lceil j \rceil_N]$. Then the value of w is *true*, as r is a successful read. So w is not an initial write (these are all assumed to be of value *false*). Hence w belongs to some unique Kan event K, and we define

$$\Gamma(L_j) = K.$$

THEOREM 3.2. *Γ is one-to-one.*

Proof. Suppose that $L_{j_1} \neq L_{j_2}$ are successful Lo events, and we prove that $\Gamma(L_{j_1}) \neq \Gamma(L_{j_2})$. Let r_1 and r_2 be the successful reads in L_{j_1} and L_{j_2} respectively. Then, by definition of Γ, $\omega(r_1) \in \Gamma(L_{j_1})$, and $\omega(r_2) \in \Gamma(L_{j_2})$. If we prove that $\omega(r_1) \neq \omega(r_2)$, then $\Gamma(L_{j_1}) \neq \Gamma(L_{j_2})$ follows since every Kan event contains a single write. Assume for a contradiction that $\omega(r_1) = \omega(r_2) = w$. Then w is a write on some $A[k]$ and hence both r_1 and r_2 are reads of $A[k]$. This implies that $\lceil j_1 \rceil_N = \lceil j_2 \rceil_N = k$. Since $L_{j_1} \neq L_{j_2}$, $j_1 \neq j_2$. Assume for example that $j_1 < j_2$ and then $L_{j_1} \prec L_{j_2}$. Let t be the write in L_{j_1} of value *false* on register $A[k]$. We thus have

$$w \prec r_1 \prec t \prec r_2.$$

But this and $w = \omega(r_2)$ contradicts the requirement on the return function $\omega(r_2)$ which is supposed to give the rightmost write on $A[k]$!

3.1. Formalizing the proof. The development of the *KanGaroo/LoGaroo* Protocol and the proof of Theorem 3.1 were rather informal, and in particular the flowcharts were not mentioned, so that the definition (in Chapter 3) of the protocol semantics was not used explicitly. This should not be taken as a criticism of that proof, as semi-formal outlines do have their place and most workers would probably rarely need completely formal proof. A semi-formal proof is acceptable, in my opinion, if it meets the following requirements:

(1) Most readers (from the intended readership) would agree that the proof is convincing and sufficiently clear.
(2) The proof can be expanded into a formal one.

The aim of this section is to show through an example what I mean by expanding an informal proof. Special attention is given to the definition of the higher-level events. The presentation will be brief though, because a full presentation would necessitate repeating the previous section. For simplicity, we assume a single *KanGaroo*, that is $p_0 = 1$.

Notice that the read/write instructions are binary

$$Read(a, x), \; Write(a, v)$$

rather than the unary register operations in Peterson's Protocol. The first parameter, a, refers to one of the registers $A[0], \ldots, A[N-1]$, and the second parameter (or variable) is boolean. Formally, A is a function defined on *TagNumber* such that $A[n]$ is a register. So $A : TagNumber \to registers$. We write $A[n]$ rather than $A(n)$, because this is the traditional form.

We recall from Chapter 3 that the first step is to define the background structure. It is particularly simple in our case. Its sorts are the natural numbers, the tag numbers (*TagNumbers* are the natural numbers t with $0 \leq t < N$), the values *true/false*, and finally the names of the registers. Two functions on natural numbers are defined. The successor function $+1$, and the mode N function $\lceil \cdot \rceil_N$ which gives to each $i \in N$ the tag number $\lceil i \rceil_N = i \pmod N$. Then A is a function defined on *TagNumber* such that $A[i]$ is a register.

Then the communication devices and the external operations must be specified. These are the register $A[i]$, $0 \leq i < N$, with their read/write operations, and the obtain_a_tag operation.

The flowcharts are obtained from the protocol. The combined higher-level and explicit flowcharts are drawn in figures 5.19 and 5.20.

The definition of the states and of the transitions of the flowcharts is left as an exercise. Let $T = (\sigma_K, \sigma_L)$ be a history of the explicit E_K/E_L flowchart. So $\sigma_K(0) = \sigma_L(0)$ is a global state (located at the "concurrently do" node 0), and $\langle \sigma_K(i), \sigma_K(i+1) \rangle$ (as well as $\langle \sigma_L(i), \sigma_L(i+1) \rangle$) are transitions in E_K (respectively E_L) whenever defined.

T induces a history $T \mid H_K/H_L = (\tau_K, \tau_L)$ on the combined H_K/H_L flowchart. Essentially, τ_K (for example) is a subsequence of σ_K obtained by taking those states of σ_K that can be viewed as states of H_K. That is, those states $\sigma_K(i) = S$ with $S(C) \in \{0, 1, 2, 3\}$. With one exception: If $u = \sigma_K(i)$ is the unreachable state u with $u(C) = 2.2$ (corresponding to a non-terminating execution of obtain_a_tag), then the induced H_K state is the unreachable state U with $U(C) = 3$

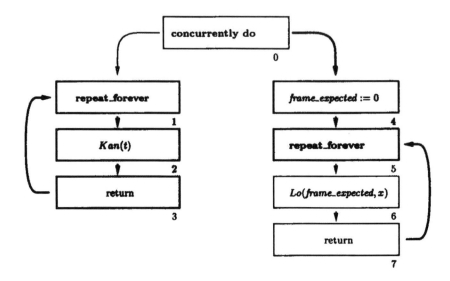

FIGURE 5.19. The combined higher-level H_K/H_L flowcharts.

(corresponding to a non-terminating call of the *Kan* procedure).

A step is a representation of an execution of a transition. $\langle \sigma_K, i \rangle$ is the ith step in σ_K, representing execution of transition $\langle \sigma_K(i), \sigma_K(i+1) \rangle$ (when $i+1$ is in the domain of σ_K). Similarly, $\langle \sigma_L, i \rangle$ is the ith step in σ_L, representing a transition in E_L. The higher-level steps are of the form $\langle \tau_K, i \rangle$ and $\langle \tau_L, i \rangle$ representing H_K and H_L transitions.

The substep relation relates steps of σ_K and steps of τ_K (and similarly steps of σ_L and steps of τ_L). By flattening the two-level history structure $(\sigma_K, \sigma_L, \tau_K, \tau_L)$, a two-level history model is obtained, which we still denote by T. T is rich enough to allow definition of sets of steps.

The set of all steps between an "activate *Kan*" and the corresponding "conclude *Kan*" step in σ_K are collected into a set which is said to be "an execution of *Kan*". Similarly, executions of *Lo* are defined as collections of steps of E_L. Each execution of *Kan* is identified with a higher level τ_K step (and similarly each execution of *Lo* is identified with a higher level τ_L step). Specifically, a higher-level H_K step (or H_L step) is identified with the set of all its substeps in σ_K (or in σ_L).

An execution of *Lo* is said to be successful if it contains a substep of reading $A[\lceil frame\_expected \rceil_N]$ with value *true*. That is, a substep $\langle \sigma_L, i \rangle$ with $\sigma_L(i)(C) = 6.2$ such that $\sigma_L(i+1)(frame\_expected) = true$. This substep is said to be "successful".

Since the reading steps that precede the successful step are not used in the proof, it is convenient to form higher-level events that do not include them, and we thus define a "*Lo* event" as a collection consisting of only a successful step and the subsequent writing step (corresponding to node 6.4).

Executions of the *Kan* procedure can either be terminating or non-terminating, depending on whether the obtain_a_tag step is terminating or not. *Kan* events

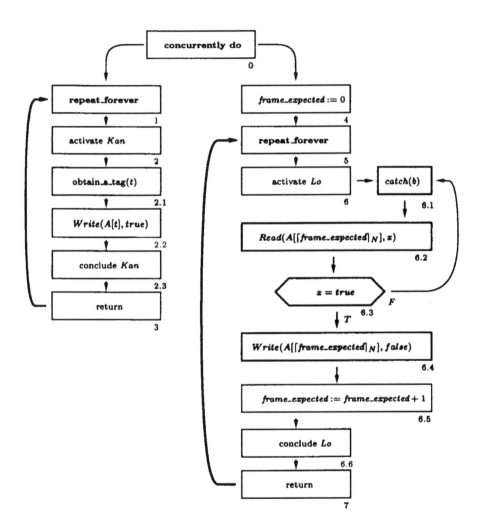

FIGURE 5.20. The combined explicit E_K/E_L flowcharts.

are defined accordingly.

(1) A terminating Kan event is a subset of a terminating Kan execution that contains only two steps: the obtain_a_tag step and the write step (corresponding to node 2.2). That is, the activation and conclusion steps are omitted. (Though they were not used in our arguments, they were instrumental in defining the higher level executions, just as brackets are used to determine the extension of a formula.)

(2) A non successful Kan event is a singleton containing a non-terminating obtain_a_tag step.

The name of the procedure, Kan, is used as a predicate, and we write $Kan(X)$ if $X = \{o, w\}$ (or $X = \{o\}$) is a successful Kan event (non successful Kan event respectively), where o is the obtain_a_tag step and w the write step. Similarly Lo is a predicate and $Lo(X)$ is defined.

If $Kan(X)$ and X is terminating, then $tag(X)$ denotes the tag number obtained by o. More formally, if event o represent the transition $\langle \sigma_K(i), \sigma_K(i+1) \rangle$, then $tag(X) = \sigma_K(i+1)(t)$.

The ordering $<_{steps}$ on the steps induces an ordering, still denoted $<_{steps}$, on the higher-level steps. If A and B are sets of steps, then we write $A <_{steps} B$ iff $\forall s \in A \ \forall t \in B \ (s <_{steps} t)$.

We note now some properties of the protocol's executions that hold in our history model H. These properties rely on the internal semantics alone. Afterwards we shall consider the external semantics, which depends on the registers, and conclude the desirable correctness properties.

(1) The Lo events in H are serially ordered. We let $\langle L_j \mid j \in J_0 \rangle$ be the enumeration of the Lo events in increasing order: $L_0 <_{steps} L_1 <_{steps} \cdots$.

(2) For every $j \in J_0$, $frame\_expected(L_j) = j$.

(3) If $j \in J_0$ and L_j is successful, then $L_j = \{r, w\}$ with $r <_{steps} w$, where r the successful read, is a read of register $A[\lceil j \rceil_N]$ returning $true$, and w writes $false$ on that register. If L_j is not successful, then $j = \max(J_0)$, and L_j consists of an infinite set of reads of register $A[\lceil j \rceil_N]$, all returning $false$.

(4) Every Kan event X with $tag(X) = k$ contains a write of $true$ on $A[k]$. Every write step on register $A[k]$ of value $true$ is in some unique Kan event X such that $tag(X) = k$ (mod N). The initial write on $A[k]$ is of value $false$, and any other write of $false$ is in some L_j with $j = k$ (mod N).

Now consider a system execution S that extends H, and such that the registers $A[i]$, $0 \le i < N$, are serial (with a return function ω). The lower-level events of S are the steps of H, and the higher-level events are the Kan/Lo events as defined above, viewed as sets of events.

We shall define a function, Φ, on the successful Lo events. To define $\Phi(L)$ (where $Lo(L) \wedge successful(L)$ holds) let $j = frame\_expected(L)$ and let $r \in L$ be the successful read in L. r is a read of $A[\lceil j \rceil_N]$ and $Value(r) = Value(\omega(r)) = true$. By property (4), $\omega(r)$ is necessarily in some unique Kan event K with $tag(K) = \lceil j \rceil_N$. We define then

$$\Phi(L) = K.$$

If $L = L_{j_0}$ is a non-successful Lo event, then $L = \{r_0, r_1, \ldots\}$ is an infinite set of read events, all of register $A[k]$ where $k = \lceil j \rceil_N$, and such that $Value(r_i) = false$ for all i. Thus $\omega(r_i)$ are writes of value $false$ on $A[k]$. Hence $\omega(r_i)$ is either the initial write on $A[k]$, or else $\omega(r_i) \in L_\ell$ for some ℓ with $\ell = k \pmod{N}$. Suppose that the second alternative holds, and then:

(1) $\ell < j_0$ (since $\omega(r_i) \prec r_i$).
(2) $\lceil \ell \rceil_N = k$.
(3) Hence $j_0 \geq N$ and $\ell = j_0 - N$.
   Proof: Since $0 \leq \ell < j_0$ and $\ell = j_0 \pmod{N}$, $j_0 \geq N$ and $\ell \leq j_0 - N$ follow. But if $\ell < j_0 - N$, then the inequalities $\ell < j_0 - N < j_0$, and the fact that $L_{j_0 - N}$ contains a write on $A[k]$, imply that $\omega(r_i)$ is not the rightmost write on $A[k]$ preceding r_i.

THEOREM 3.3. *Suppose that L_{j_0} is not successful, and let K be some terminating (successful) Kan event with $tag(K) = \lceil j_0 \rceil_N$. Then $j_0 \geq N$ and it is not the case that $L_{j_0 - N} \prec K$.*

Proof. By Lamport's finiteness condition applied to K, there is only a finite number of events x for which $K \prec x$ does not hold. So, since L_{j_0} is infinite, there is some $r \in L_{j_0}$ with $K \prec r$. r is a read of $A[k]$, where $k = \lceil j_0 \rceil_N$, returning $false$. Since $tag(K) = \lceil j_0 \rceil_N$, K contains a write w on $A[k]$ of value $true$, and hence $w \prec \omega(r)$. So $\omega(r)$ is not the initial write. Hence, by our above observation, $j_0 \geq N$ and $\omega(r)$ is in $L_{j_0 - N}$. So $L_{j_0 - N} \prec K$ is impossible (since otherwise, as $\omega(r) \in L_{j_0 - N}$, $\omega(r) \prec w \prec r$). $\square$

4. The Aimless Protocol

Finding an appropriate degree of formality (or informality) is very important for those who appreciate the usage of proof to clarify ideas and concepts. The proofs in the second and third part of the book are quite informal in comparison with those that were described so far, and no flowcharts, states, or histories are mentioned. The aim of this last example (the Aimless Protocol) is to describe a degree of formality which in my experience is most useful. The informality consists in going directly from the protocol to its executions, bypassing the exact definition of semantics of Chapter 3. This makes the proofs accessible to a wider readership. The only prerequisite for reading such proofs is the notion of system execution, as described for example in Chapter 2.

The *Aimless* Protocol[1] is in Figure 5.21. Registers R_0 and R_1 are serial. We assume that the initial value of both registers is 0, and we let w_0 and w_1 be the initializing writes on R_0 and R_1. The protocol can be simply described in the following words. *Copier* reads R_1 and writes the value obtained on R_0, and *Changer* reads R_0 and writes on R_1 the opposite value. (The values of variables s, t are in $\{0, 1\}$.)

---

[1]The Aimless Protocol was investigated by Dolev et al. [15], by Peterson [26], and it also appears (although indirectly) in Bloom's two writers protocol [11].

procedure *COPY* var s : 0,1 ; (1) read$R_1(s)$; (2) write$R_0(s)$.	procedure *CHANGE* var t : 0,1 ; (1) read$R_0(t)$; (2) write$R_1(1-t)$.
procedure *Copier* **repeat_forever** *COPY*.	procedure *Changer* **repeat_forever** *CHANGE*.

> **protocol *Aimless***
> **concurrently** do *Copier*, *Changer* od.

FIGURE 5.21. *COPY* and *CHANGE* with serial registers.

As is our custom, we use the same name *CHANGE* (*COPY*) to denote both a procedure (i.e., the instruction written in figure 5.21) and its higher-level executions. The following two paragraphs describe informally the protocol and what we intend to prove about it.

A single execution of the *CHANGE/COPY* procedure (in a system execution that describes an execution of the *Aimless* Protocol) is called a *CHANGE/COPY operation execution*. Let us look first at a particularly simple type of system executions in which no interleaving of operations exist: The *CHANGE* and *COPY* operations are repeatedly executed, and for any two such executions X and Y, either X precedes Y or Y precedes X. We call such executions "atomic executions" because the read followed by the write events corresponding to a single operation execution are thought of as being executed at once. These system executions are very simple to describe and to understand. If we color each operation execution with 0 or 1, according as to the value written on the register, then the blocks of 0's and 1's are arranged in alternations. A block of color $c = 0, 1$ of *Copier* is followed by a block of *Changer* in the complementary color $1-c$. However, if the operations of the *Changer* and *Copier* are interleaved, then a more complex pattern results and our aim is to understand it.

Specifically, we will try to prove the following property of general executions. A block of *Copier* (or Changer) is defined to be a maximal consecutive sequence of *COPY* (or *CHANGE*) operation executions that have the same color. It is not clear what is the relation between *Changer* and *Copier* blocks. We want to prove that for any two operation executions X and Y in the same block, if X reads the register written by A and Y reads the register written by B, then A and B are in the same block as well. In the simple case of atomic executions, X and Y are in the same block iff $A = B$, but in general runs this is not necessarily true.

When the *Aimless* Protocol is executed, *Copier* and *Changer* concurrently and repeatedly execute their procedures, *COPY* and *CHANGE*. The resulting run is modeled by a system execution S with a set of events E and a precedence relation $\prec$. E contains all the read/write events on registers R_0 and R_1, as well as the higher-level events defined below.

DEFINITION 4.1 (OF PREDICATES *COPY*, *CHANGE*, *Copier* , *Changer*). *Each higher-level event is formed by grouping events in E that belong to a single com-*

plete execution of either the COPY *or the* CHANGE *procedure. There are thus two kinds of higher-level events in our example: a* COPY *event which represents an execution of the* COPY *procedure, and a* CHANGE *event which represents an execution of the* CHANGE *procedure. If X is a* COPY *higher-level event, then $X = \{r, w\}$ contains two events: r is a read of R_1 and w is a write on R_0. These are successive events in* Copier *and* $Value(r) = Value(w)$.

If X is a CHANGE *event, then X is again a set of two lower–level events: a read and a write events, but now $X = \{r, w\}$ where r is a read of R_0, w a write on R_1, and $Value(w) = 1 - Value(r)$.*

E *is partitioned into two sets by the predicates* Copier *and* Changer*. For example,* Copier(e) *holds when e is an execution of an instruction from* COPY *or is a* COPY *higher level event, or is the initial write w_0 on R_0.* Changer(e) *is similarly defined.*

The first event in *Changer* is an initializing event, w_1, and it is followed by CHANGE events (operation execution). The first event in *Copier* is an initializing event w_0 followed by COPY operation executions. The two *Initial* events precede all other events, but CHANGE and COPY events may overlap in S (i. e., be $\prec$ incomparable).

The collection of these higher-level events, together with the precedence relation $\prec$, and the functions and relations that will be defined later, form the structure enabling the specification and correctness proof of the protocol.

Recall that $\prec$ also relates higher and lower-level events: for example $X \prec a$ for $X \in H$ and $a \in E$ holds if and only if $\forall x \in X (x \prec a)$.

4.1. Higher-level relations. Let system execution S be an execution of the Aimless Protocol, as described in the previous subsection. Let H denotes the set of higher-level events. For uniformity, it is convenient to treat the initial events as higher-level events, and therefore we write $I_0 = \{w_0\}$ and $I_1 = \{w_1\}$. So $I_0, I_1 \in H$.

DEFINITION 4.2.   (1) *The function "color" is defined on the events in H. If X is a* COPY/CHANGE *event, then* color(X) *is the value (0 or 1) of variable s, or t, written onto register R_0 (if X is a* COPY*) or R_1 (if X is a* CHANGE*). We assume that* color$(I) = 0$ *for an Initialization event $I = I_0, I_1$.*

(2) *The function Φ is defined on the* COPY *and* CHANGE *events in H by*

$$\Phi(U) = V$$

iff for the (only one) read event $r \in U$, $\omega(r) \in V$. When the write $\omega(r)$ is the initial write onto the register, then $\Phi(U) = I_0, I_1$ and we say that U "observes" the Initialization. So Φ takes COPY *events into* Changer *events, and* CHANGE *events into* Copier *events.*

I suggest that the reader accompany this definition and the following proofs with illustrations, just as she would do in geometry. I use the following conventions, but the reader is free of course to choose her own. Time flows on horizontal lines from left to right. Each line is reserved for a single linear process; so that

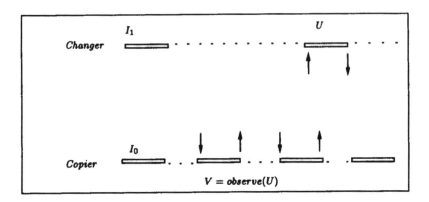

FIGURE 5.22. An illustration for Lemma 4.3.

we have two lines, the bottom line for *Copier* and the upper for *Changer*. Lower-level read/write events are represented as arrows reaching these lines (for reads) or emanating from the lines (for writes). Values are represented by zeros and ones. Narrow rectangles represent operation executions.

The following lemma lists some properties of these functions that are quite easy to prove intuitively.

LEMMA 4.3. *The function, Φ, and color satisfy the following for any CHANGE, COPY event U:*

(1) (a) $\Phi(U) \prec U$,
    (b) *If $V = \Phi(U)$ and V' is in the same process as V (Changer or Copier) then $V \prec V' \prec U$ is impossible. (This property of Φ is called* regularity.*)*

(2) *If U is a CHANGE event then $\Phi(U)$ is in Copier and*

$$color(U) \neq color(\Phi(U)).$$

(3) *If U is a COPY then $\Phi(U)$ is in Changer and*

$$color(U) = color(\gamma(U)).$$

The proof is left as an exercise, except for (1)(a) for which we outline a proof. Assume that $\Phi(U) = V$, and we prove that $V \prec U$. An illustration of this situation when U is a *CHANGE* event is in figure 5.22.

According to the definition of $\prec$ on higher-level events, we must prove that

$$\forall v \in V \ \forall u \in U (v \prec u).$$

By definition, $\Phi(U) = V$ means that if r is the read event in U corresponding to line 1, then $\omega(r)$ is in V. Since $\omega(r)$ is a write, it follows that it is either the initial write on its register, or else V is an execution of the procedure, and then $\omega(r)$ is the last event in V. As $\omega(r)$ is the last event of V and r is the first event of U, $V \prec U$ follows from $\omega(r) \prec r$.

Lemma 4.3 is a pivot in the formal development of our discussion: From now on no reference to the protocol's text and to the lower-level events will ever be made, and only properties stated in this lemma will be used. We crystallize this in the following definition.

DEFINITION 4.4. *Let $\mathcal{H}$ be a system execution for the following signature:*

(1) *Two sorts: events and colors (where color $= \{0, 1\}$).*
(2) *Five unary predicates are defined on the events: Changer, Copier, CHANGE, COPY, Initial.*
(3) *Two unary functions are defined on the events: Φ and color. Function color assumes values in sort "color", and Φ is from events to events. In fact, Φ takes CHANGE events into Copier, and COPY events into Changer.*

We say that $\mathcal{H}$ is a "Changer/Copier" system execution if and only if the following hold:

(1) *Every event falls either in Changer or in Copier, but not in both. Changer and Copier are serial. That is (for Changer, for example)*

$$\forall X, Y(Changer(X) \wedge Changer(Y) \wedge X \neq Y \rightarrow X \prec Y \vee Y \prec X).$$

(2) *There is a single event I_1 in Changer that is an initial event (i.e., $Initial(I_1)$ holds), and a single initial event I_0 in Copier. These initial events precede any other event.*
(3) *Changer consists of its initial event and the CHANGE events, and similarly Copier consists of its initial event and the COPY events.*
(4) *The main requirement is that the properties established in Lemma 4.3 hold.*

LEMMA 4.5. *Suppose that $\mathcal{H}$ is a Changer/Copier system execution. Then the following hold in $\mathcal{H}$.*

(1) *For any events $U_1 \preceq U_2$ that are both CHANGE or both COPY events,*

$$\Phi(U_1) \preceq \Phi(U_2).$$

(2) *For every CHANGE/COPY event U, if $\neg Initial(\Phi(U))$ (and then $\Phi(\Phi(U))$ is defined), then*

$$(9) \qquad\qquad \Phi(\Phi(U)) \prec U$$

and

$$color(U) \neq color(\Phi(\Phi(U))).$$

To see the first item, assume $U_1 \preceq U_2$ and put $W_i = \Phi(U_i)$, for $i = 1, 2$. If $W_1 \preceq W_2$ does not hold, then $W_2 \prec W_1$ by the seriality of Changer and Copier. But $W_1 \prec U_1$ by Lemma 4.3.(1)(a). So $W_2 \prec W_1 \prec U_1 \preceq U_2$ follows in contradiction to the regularity of Φ (Lemma 4.3.(1)(b)).

Equation 9 follows by a double application of property (1)(a). Given any U, $\Phi(U) \prec U$ by 1(a), and if $\Phi(U)$ is not the initial event, then Φ can be applied again, and $\Phi(\Phi(U)) \prec \Phi(U)$. Hence by transitivity of $\prec$ the result follows.

As for the color difference, it follows from properties (2) and (3) of Lemma 4.3 □

The following notion *Dom* turns out to be central in our proof.

DEFINITION 4.6.   (1) *For every CHANGE event $V \in H$, define $Dom(V) \subset$*
Copier as follows. Let $U_0 = \Phi(V)$. For any Copier event $U \in H$, we let
$U \in Dom(V)$ iff: $U = U_0$ or else the following holds
   (a) $U_0 \prec U$, *and*
   (b) $color(U) = color(U_0)$ *and*
   (c) $\Phi(U) \prec V$.
   (2) *Similarly, for every COPY event, $V \in H$, define $Dom(V) \subset$ Changer*
as follows. Let $U_0 = \Phi(V)$. For any Changer event $U \in H$, we let
$U \in Dom(V)$ iff: $U = U_0$ or the following holds
   (a) $U_0 \prec U$, *and*
   (b) $color(U) = color(U_0)$ *and*
   (c) $\Phi(U) \prec V$.

Recall that if $<$ is any ordering, then a convex set is a set K such that if
$a < b < c$ and $a, c \in K$, then necessarily $b \in K$.

THEOREM 4.7. *Let $\mathcal{H}$ be a Changer/Copier system execution. For any CHANGE,*
COPY event, V, $Dom(V)$ is convex. (I.e., for example, if U_0, U_1, U_2 are in Copier
and $U_0 \prec U_1 \prec U_2$ and $U_0, U_2 \in Dom(V)$, then $U_1 \in Dom(V)$ as well.)

Proof. Intuitively the proof is by induction on $Right\_End(V)$, namely on the
last lower-level event in V. However, since there are no lower-level events here
we argue as follows (see page 29 for a general discussion of inductive proofs).

Any *CHANGE/COPY* operation execution V is terminating, and the finiteness
condition of Lamport (Definition 1.5) implies that $P(V) = \{X \mid \neg(V \prec X)\}$ is
finite. Thus, if the theorem does not hold, there is a least counter-example V
(that is, one with a minimal number of events in $P(V)$). Assume that V is a
CHANGE event (the proof is symmetric in case V is a *COPY* and is left to the
reader). So,

   (1) $Dom(V)$ is not convex, that is for some *Copier* events $U_0 \prec U_1 \prec U_2$,
   $U_i \in Dom(V)$ for $i = 0, 2$ but not for $i = 1$.
   (2) If $|P(X)| < |P(V)|$ then X is not a counterexample to the theorem.

Since, by definition, $\Phi(V)$ is the $\prec$-first event in $Dom(V)$, $\Phi(V) \prec U_1$ and there
is no loss of generality in assuming that $U_0 = \Phi(V)$. Put

$$c = color(U_0),$$

then $c = color(U_2)$ (since $U_2 \in DOM(V)$). As V is a *CHANGE* event,

$$\bar{c} = 1 - c = color(V).$$

U_1 and U_2 are *COPY* events, because the *Initial* events are not preceded by
any event. Hence they are in the domain of the function Φ. Let, for $i = 1, 2$,

$$W_i = \Phi(U_i).$$

Then:

   (1) $color(U_i) = color(W_i)$, for $i = 1, 2$, since U_i are *COPY* events. In particular, $c = color(U_2) = color(W_2)$.
   (2) $W_1 \preceq W_2$, by Lemma 4.5.

Since $U_2 \in Dom(V)$,

$$W_2 \prec V.$$

Hence $W_1 \prec V$. So U_1 satisfies (a) and (c) of the definition of $Dom(V)$ (Definition 4.2), and if $c = color(U_1)$ then $U_1 \in Dom(V)$ is the required contradiction. So assume

$$\bar{c} = 1 - c = color(U_1).$$

Hence

$$\bar{c} = color(W_1).$$

It follows from what we have proved so far that U_1 is also a counterexample to the theorem:

(1) $W_1 = \Phi(U_1)$, and
(2) $V \in Dom(U_1)$ (because $\bar{c} = color(W_1) = color(V)$ and $\Phi(V) \prec U_1$), but
(3) $W_1 \preceq W_2 \prec V$, and yet $W_2 \notin Dom(U_1)$ as $c = color(W_2)$.

However,

$$|P(U_1)| < |P(V)|$$

follows and contradicts the minimality of V. To prove this inequality, prove first that $U_2 \not\prec V$ (by assuming contrariwise that $V \prec U_2$, and then $W_2 = \Phi(U_2) \prec V$ is a contradiction). Now, since $U_1 \prec U_2$, $U_1 \prec U_2 \not\prec V$, and this implies $P(U_1) \subseteq P(V)$ (for if $V \prec X$ then $U_1 \prec X$ by the Russell–Wiener property, and hence $X \notin P(V)$ implies $X \notin P(U_1)$). Yet $U_2 \in P(V) \setminus P(U_1)$. $\square$

We can now approach our aim of understanding the behavior of the Aimless Protocol.

DEFINITION 4.8. *Let $\mathcal{H}$ be a Changer/Copier system execution. Suppose that X and Y are both CHANGE (or both COPY) operation executions. We say that X and Y are "in the same block" iff for some color $c \in \{0,1\}$ $c = color(Z)$ for every CHANGE (respectively COPY) operation Z such that $X \preceq Z \preceq Y$. (in particular $c = color(X) = color(Y)$.)*

The proof of the following lemma is very simple now.

LEMMA 4.9. *(1) Being in the same block is an equivalence relation and each equivalence class is convex.*

(2) For any V, all events in $Dom(V)$ are in the same block. (Use Theorem 4.7)

(3) If X and Y are not in the same block, then neither are $\Phi(X)$ and $\Phi(Y)$ in the same block.

COROLLARY 4.10. *Let $\mathcal{H}$ be a Changer/Copier system execution, and let X and Y be in the same block. Then $A = \Phi(X)$ and $B = \Phi(Y)$ are in the same block as well.*

Proof. Let X and Y be in the same block, and assume (without loss of generality) that $X \prec Y$. Let c be the common color of X and Y. Then $color(A) = color(B)$, because if X, Y are in *Copier* then $color(A) = color(B) = c$, and if both are in *Changer* then $color(A) = color(B) = \bar{c}$. By Lemma 4.5(1), $A = \Phi(X) \preceq B = \Phi(Y)$.

We must prove that if C is an operation execution in the same process with A and B (a *Changer* if X and Y are *Copier*, and a *Copier* otherwise, and if $A \prec C \prec B$, then $color(A) = color(C)$ $(= color(B))$. We will obtain this result by proving that $A, B \in Dom(X)$ and invoking Theorem 4.7 to conclude that A and B are in the same block.

Since $A = \Phi(X)$, clearly $A \in Dom(X)$. Now, since A and B have the same color, we only have to prove that $\Phi(B) \prec X$ in order to conclude that $B \in Dom(X)$ as well. But if this is not the case, then $X \preceq \Phi(B)$ by the seriality of the process. Yet $\Phi(B) \prec Y$ and the color of $\Phi(B)$ is different from that of Y (both by Lemma 4.5(2)). Thus $\Phi(B)$ is between X and Y, and this shows that X and Y are not in the same block! $\square$

The meaning of this corollary is that Φ induces a map from the blocks of one process into the blocks of the other. Specifically, given any block B, look at all the events of the form $\Phi(X)$ for $X \in B$; then they all belong to a unique block which we may denote $\Phi(B)$. We leave it to the reader to prove that Φ is order preserving: If $W_1 \prec W_2$ are *CHANGE* (or *COPY*) events in different blocks, then $\Phi(W_1) \prec \Phi(W_2)$.

EXERCISE 4.11. *Let $\mathcal{H}$ be a Changer/Copier system execution such that every block contains exactly one operation execution. Suppose that the first CHANGE operation observes the first COPY operation. Prove that the the COPY/CHANGE operation executions alternate: There is no overlap between operations, and after any COPY (CHANGE) operation there is an immediate successor which is a CHANGE (COPY) operation.*

EXERCISE 4.12 (PETERSON–FISCHER MUTUAL EXCLUSION). *The protocol in Figure 5.23 ensures not only the mutual exclusion property, but in fact the critical sections of P_0 and P_1 alternate with P_0 executing its critical section first. It is a variant of a protocol of Peterson and Fischer [26]. Registers R_0 and R_1 are serial, assume values in $0..1$ and their initial value is 0. (The initial values of all variables is 0 as well.) Notice the similarity between this and the Aimless Protocol. P_0 resembles the Copier and P_1 the Changer. Hence we can use the results obtained for the Aimless Protocol.*

5. Local registers

The distinction between internal and external semantics is a corner stone of the approach described here. States and histories specify internal semantics, while events and system executions are for external semantics. This distinction has lead to a strict notational policy of separating assignments (which are internal operations) from external operations such as reading and writing on registers.

procedure E var in, out: 0..1;	procedure D var in, out: 0..1;
(1) while $in \neq out$ do read$R_1(out)$; (2) Critical Section$_0$ (3) $in := 1 - out$; (4) write$R_0(in)$	(1) while $in=out$ do read$R_0(in)$; (2) Critical Section$_1$ (3) $out := in$; (4) write$R_1(out)$
procedure P_0 **repeat_forever** E.	procedure P_1 **repeat_forever** D.

FIGURE 5.23. Aimless Mutual Exclusion of Peterson and Fischer. The initial values of all registers and variables is 0.

procedure Kan var t : *TagNumber*; **begin**	procedure Lo var *frame_expected*: Natural number; **begin**
(1) obtain_a_tag(t); (2) $A[t] :=$ **true** **end**	(1) **repeat** $\quad$ *wait-until* $A[\lceil frame\_expected \rceil_N]$; (2) $A[\lceil frame\_expected \rceil_N] :=$ **false**; (3) *frame_expected* := *frame_expected* + 1 **end**

FIGURE 5.24. Alternative Kan/Lo procedures.

We use $x := \tau$ only for assignments—x is a variable and τ an expression. When dealing with registers, most workers would probably prefer

$$A[t] := true$$

to our (somewhat cumbersome)

$$Write(A[t], true)$$

and use

$$x := A[k]$$

rather than

$$Read(A[k], x).$$

They would write the Kan and Lo procedures as in Figure 5.

I cannot deny that this liberal usage of := has its merits, and in some cases I will also use the $x := A$ notation even when A is a register. In particular when some process is the unique writer on A and all others can only read it. In this case the writing process may consider A to be a a local variable rather than a communication register, and insisting on read/write instructions would be too pedantic, so it seems.

DEFINITION 5.1 (LOCAL REGISTER). *A single-writer register is called a local register when the processes can treat it as a variable and use assignment rather than read/write instructions.*

PART 2

Shared-variable communication

The theory and concepts of the first part are used here to discuss shared-variable communication with an emphasis on the Consumer/Producer problem.

As described in any operating system text book (e.g., Silberschatz et al. [28]) the *Producer/Consumer* problem is one of coordinating two processes that operate concurrently and at variable speeds: the Producer repeatedly passes data items to the Consumer, and the question is how to regulate the flow of information most efficiently. Without such regulation the Producer may overwhelm the Consumer with data, before the latter can retrieve and consume it. A possible solution is to force the Producer to wait after each item for an acknowledgment from the Consumer, indicating that it may continue with the next item. This however slows down the processes to an unacceptable degree. The basic idea of the solution is just like the one for regulating water flow: to create a reservoir, which is an array of buffers that can retain data items until they are needed. This is investigated in the following two chapters.

Chapter 6 defines the problem formally and specifies buffers and semaphores which are used in its solution. Chapter 7 describes a circular buffer protocol.

6

On the Producer/Consumer problem: buffers and semaphores

The first section explains and specifies the Producer/Consumer problem. The second section specifies buffers. The third section discusses semaphores, and shows the equivalence between two specifications. All of this is used in the fourth section to prove the correctness of the Dijkstra Producer/Consumer protocol.

1. The Producer/Consumer problem

Consider two processes called *Consumer* and *Producer* that continuously repeat the following operations: The producer produces an item and hands it over to the consumer who takes and consumes it. For example, the producer may be a radio scanner and the consumer an analyzer of the data obtained. The scanner directs its antennas to outer space and scans a range of frequencies; it listens for a very short period to each frequency in its range (one frequency after the other) and hands out the raw data obtained to the analyzer (the consumer) which checks if this is just noise or some interesting information. In the latter case it may alert a more sensitive receiver to tune in to the interesting frequency for a longer period. This producer-consumer team must act in good and swift coordination because any lost item may be important.

In general, the operations of producing an item and consuming it are of variable and unequal lengths. A policy of strict coordination in which the producer waits until the consumer is ready for the next item and then directly transfers it to the consumer may slow down the system to an unacceptable degree. A buffer is therefore useful here, using storage space to save time. Even while the consumer is busy with analyzing the previous data, the producer can load the buffer and immediately turn to producing the next item.

A *buffer* consists of a finite array of buffer cells, each carrying an item. The producer loads these cells and the consumer unloads them. The protocol tells each process which buffer to approach, so that the overall operation is an orderly and swift transfer of the items from *Producer* to *Consumer*. "Orderly" means that the first item produced is also the first consumed, and so on.

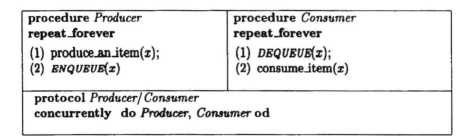

procedure *Producer*	procedure *Consumer*
repeat_forever	**repeat_forever**
(1) produce_an_item(x);	(1) *DEQUEUE*(x);
(2) *ENQUEUE*(x)	(2) consume_item(x)

protocol *Producer/Consumer*
concurrently   **do** *Producer, Consumer* **od**

FIGURE 6.25. The generic *Producer/Consumer* protocol

The simpler protocols use a single cell for buffering. The aim of the protocol in this case is to ensure that the consumer and producer never approach that cell simultaneously, but rather alternate, in order, with the producer having the first turn. More complex protocols employ an array of buffer cells, which has the advantage of allowing the producer and consumer to approach different buffer cells simultaneously. In this chapter, only single cell protocols are described, and in the following chapter multi-cell, circular buffer protocols.

All of our *Producer/Consumer* protocols have the form of Figure 6.25, and the differences between the different protocols presented in this and the subsequent chapter lie only in the procedures that implement the operations *ENQUEUE*(x) and *DEQUEUE*(x). In this section *ENQUEUE* and *DEQUEUE* are considered as external operations, and we do not explicate their implementations. *ENQUEUE* has a single value parameter and *DEQUEUE* a single variable parameter, both of type *Item* which is the users' data type. Line 1 in Producer and line 2 in Consumer denote some external operation left unspecified: x is a variable parameter in "produce_an_item" returning a value in x, and it is a value parameter in "consume_item". For simplicity, we assume that these are always terminating operations.

To execute operations *ENQUEUE/DEQUEUE* the system invokes procedures (with the same names) that implement the operations, and the correctness of the protocols means that whenever the protocol's procedures are used to implement these operations the resulting transfer of items from the producer to the consumer is in order. To explicate this requirement formally, a first-order language L is defined below in which the requirements can be specified.

The signature of L contains three sorts: *Events*, *Item*, and N.

(1) Two standard predicates are defined on the events: *Terminating* (a unary predicate) and $\prec$ (the binary precedence relation).

(2) There are four additional unary predicates on the events: *produce_event*, *ENQ_event*, *DEQ_event*, and *consume_event*. (If *produce_event*(e), then event e is supposed to represent an execution of line (1) in procedure *Producer*, and we say that e is a produce event in this case. An event e such that *ENQ_event*(e) holds represents an execution of line 2, and we say that e is an *ENQUEUE* event, etc.) Two additional predicates are definable (i.e., they need not be part of the original signature): (1) *Producer*(e) iff *produce_event*(e) or *ENQ_event*(e). (2) *Consumer*(e) iff *DEQ_event*(e) or

consume_event(e).

(3) A function *Value* assigns an item to each event. That is,

$$Value : Events \rightarrow Items.$$

(In words: if e is a produce event then *Value*(e) is the value of the item produced; if e is an *ENQUEUE* event then *Value*(e) is the item used as parameter for the invoked *ENQUEUE* procedure, etc.)

(4) Sort N (interpreted as the set of natural numbers N) comes with its standard functions and relations such as $+$ and $\leq$. We need N in order to be able to state, for example, that the value of the nth *ENQUEUE* operation is the same as the value of the nth *DEQUEUE* operation.

(5) A function *number* is defined on the *ENQUEUE/DEQUEUE* events; its range is N. If *ENQ_event*(e) and *number*(e) $= 0$, then we say that e is the first *ENQUEUE* event, etc.

Formal first-order sentences are quite difficult to read (and write), and so we often use in this book mathematical English to describe such sentences. Still, it is a good practise to workout in detail the translations of English descriptions into formal sentences in order to gain confidence in this process. We give some examples of English versus formal writing in the language L, and only then describe the specifications of the *ENQUEUE/DEQUEUE* operations. We assume that variables e, e_1, e_2 vary over events; n, k over natural numbers.

- "The function *number* is one-to-one on the *DEQUEUE* events, and it is one-to-one on the *ENQUEUE* events." The first part is formally written

$$\forall e_1, e_2(DEQ\_event(e_1) \land DEQ\_event(e_2) \land e_1 \neq e_2 \rightarrow$$
$$number(e_1) \neq number(e_2))$$

- "There is only a finite number of *ENQUEUE* events." Assuming that *number* is one-to-one, this can be expressed by saying that the range of *number* on the *ENQUEUE* events is bounded.

$$\exists k \forall e(ENQ\_event(e) \rightarrow number(e) \leq k).$$

- "The function *number* restricted to the *DEQUEUE* events is order preserving."

$$\forall e_1, e_2(DEQ\_event(e_1) \land DEQ\_event(e_2) \rightarrow$$
$$number(e_1) < number(e_2) \longleftrightarrow e_1 \prec e_2).$$

Since the natural numbers are serially ordered, this statement together with the assumption that *number* is one-to-one imply that the *DEQUEUE* events are also serially ordered.

- "*number* is an order-preserving enumeration of the *DEQUEUE* events." We have already written-up the translation of order preservation; "enumeration" means that *number* "leaves no holes". If *number*(e) $= 2$ for some *DEQUEUE* event e, for example, then there are *DEQUEUE* events e_0 and e_1 such that *number*(e_i) $= i$ (for $i = 0, 1$). So that e is indeed the third *DEQUEUE* event if *number*(e) $= 2$. Thus "enumeration" means that the

range of *number* is an initial segment of N (which can be finite of all of
N).

$$\forall e \forall k (DEQ\_event(e) \wedge k < number(e) \rightarrow$$
$$\exists e'(DEQ\_event(e') \wedge number(e') = k).$$

- "The third *DEQUEUE* event is terminating." Together with the statement
  that *number* is an enumeration, this can be translated as

$$\exists e(DEQ\_event(e) \wedge number(e) = 2 \wedge Terminating(e))$$

What are the desirable properties of the *ENQUEUE/DEQUEUE* events? These
are of two categories: safety and liveness. (For a general, formal definition of
these categories see Alpern and Schneider [7].) The *safety property* says that
the *n*-th *DEQUEUE* execution obtains an item that was deposited by the *n*-th
ENQUEUE execution. The *liveness* property says that, under the assumption
of terminating produce/consume operations, both sets of events *DEQUEUE* and
ENQUEUE are infinite.

The specification of the *ENQUEUE/DEQUEUE* operations is given in the language L just described.

Safety requirement: For every *n* the value of the *n*th *DEQUEUE* event
is equal to the value of the *n ENQUEUE* event.

Liveness requirement: All *ENQUEUE* and *DEQUEUE* operations are terminating, and there are infinitely many such events.

---

2. Buffer cells

A buffer cell is a communication device that supports two external operations
load and unload. The buffer has a name, say *b*, that is used as parameter to
these operation. The protocol language has two external instructions $load(b, x)$
and $unload(b, x)$ to access the buffer. Parameter *b* specifies which buffer to
approach, and the second parameter *x* (of some data type) is a value parameter
in $load(b, x)$, and it is a variable in $unload(b, x)$.

An execution of $load(b, x)/unload(b, x)$ results in a load/unload event. Fixing
our buffer, we use two unary predicates on the events to denote these events:
$load(e)$ and $unload(e)$ say (respectively) that *e* is a load/unload event on buffer
b. In addition to these predicates, a function *Value* associates data values to
load/unload events is used to describe the buffer.

DEFINITION 2.1 (CONFLICT FREEDOM). *Let S be a system execution for this
signature interpreting load/unload and Value. We say that b is conflict free in
S iff the load and unload operation executions on b are serially ordered by $\prec$,
alternating load with unload and such that the first operation is a load event.*

Formally this definition can be written as follows. Let $b\_Event(x)$ be a shorthand for $load(x) \vee unload(x)$. Then buffer *b* is conflict free in S if the following
hold in S:

(1) $\forall x, y \ (b\_Event(x) \wedge b\_Event(y) \rightarrow x <_{\mathcal{H}} y \vee y \leq_{\mathcal{H}} x)$ (i.e., the events on b are serially ordered).

(2) $\forall x \neg (load(x) \wedge unload(x))$ (i.e., load/unload are disjoint predicates).

(3) $\forall x, y \ [load(x) \wedge load(y) \wedge x <_{\mathcal{H}} y \rightarrow \exists z \ (x <_{\mathcal{H}} z <_{\mathcal{H}} y \wedge unload(z)]$ (i.e., between any two loads there is an unload event).

(4) A similar sentence to the effect that between any two unload events there is a load event.

(5) There exists x such that $load(x)$ and $\forall y \ (b\_Event(y) \rightarrow x \preceq y)$. I.e., the first b-event is a load event.

DEFINITION 2.2 (SAFE BUFFER). *Let S be a collection of system executions, namely a system, for this signature. The safeness of the buffer cell b in S is the following property. For every $S \in \mathsf{S}$:*

(1) *load/unload events are always terminating.*

(2) *If b is conflict free in S then for each unload event p, if q is the last load event preceding p, then $Value(q) = Value(p)$.*

Informally, the safeness of a buffer cell is a guarantee made by the manufacturer of the cell of the following form: (1) Each load/unload operation execution is terminating. (2) If the device is used properly, namely if it is conflict free, then the value of each unload is the value of the preceding load. Thus no claim against the manufacturer of the buffer cell can be made if the malfunctioning occurs in a system execution in which the buffer cell is not conflict free.

In the protocols presented subsequently, a sequence $buf[0], \ldots, buf[N-1]$ of buffer cells is used and the safeness assumption is made for each cell. This is the 'hardware' assumption needed to ensure the correctness of the protocol.

3. Semaphores

Semaphores were introduced by E. W. Dijkstra in his 1968 paper [13], where mutual-exclusion, consumer/producer (and other) problems are discussed. We restrict our attention to unbounded semaphores (taking values in $\mathbb{N}$). Each semaphore has a name, say S, and the protocol language is supposed to have instructions $P(S)$ and $V(S)$ which we want to specify. Let's read first Dijkstra's original description:

> *Definition.* The V-operation is an operation with one argument, which must be the identification of a semaphore. (If "S1" and "S2" denote semaphores we can write "V(S1)" and "V(S2)".) Its function is to increase the value of its argument semaphore by 1; this increase is to be regarded as an indivisible operation.
>
> Note that this last sentence makes "V(S1)" inequivalent to "S1 := S1 +1". For suppose that two processes A and B both contain the statement "V(S1)" and that both should like to perform this statement at a moment when, say, "S1 = 6". Excluding interference with S1 from other processes, A and B will perform their V-operations in an unspecified order—at least: outside our control—and after the completion of the second V-operation the

final value of S1 will be = 8. If S1 had not been a semaphore but just an ordinary common integer, and if processes A and B had contained the statement "S1 := S1 +1" instead of the V-operation on S1, then the following could happen. Process A evaluates "S1 +1" and computes "7"; before effecting, however, the assignment of this new value, process B has reached the same stage and also evaluates "S1 +1", computing "7". Thereafter both processes assign the value "7" to S1, and one of the desired incrementations has been lost. The requirement of the "indivisible operation" is meant to exclude this occurrence when the V-operation is used.

Definition. The P-operation is an operation with one argument, which must be the identification of a semaphore. (If "S1" and "S2" denote semaphores we can write "P(S1)" and "P(S2)".) Its function is to decrease the value of its argument semaphore by 1 as soon as the resulting value would be non-negative. The completion of the P-operation —i.e. the decision that this is the appropriate moment to effectuate the decrease and the subsequent decrease itself—is to be regarded as an indivisible operation.

It is the P-operation which represents the potential delay, viz. when a process initiates a P-operation on a semaphore, that at that moment is =0, in that case this P-operation cannot be completed until another process has performed a V-operation on the same semaphore and has given it the value "1". At that moment more than one process may have initiated a P-operation on the very same semaphore. The clause that completion of P-operation is an indivisible action means that when the semaphore has got the value "1" only one of the initiated P-operations on it is allowed to be completed. Which one, again, is left unspecified, i.e. at least outside our control.

It is interesting to compare these informal specifications with a formal rendering in an up-to-date textbook. Z. Manna and A. Pnueli [22] use the state-transition approach to describe semaphores. I will not quote from the book because this will require explaining their notations (the interested reader will not find it difficult to access their discussion since it is in the first chapter), but instead I will try to rephrase it. Instead of the P and V operations, Manna and Pnueli use **request** and **release** operation symbols, and this makes sense, at least for me because I can never remember whether it is P or V that tries to increase the variable... The semaphore variable is an integer shared variable, say r, and it can be accessed only by **request**(r) and **release**(r) instructions. The transitions corresponding to the semaphore instructions are defined as follows:

- *request*: With the instruction

$$\ell \ : \ request(r) \ : \ \hat{\ell},$$

where ℓ and $\hat{\ell}$ are two subsequent control positions, we associate all transitions $\langle S_1, S_2 \rangle$ where S_1 and S_2 are states at control ℓ and $\hat{\ell}$ respectively (for the procedure executing this statement) such that $S_1(r) > 0$ holds, and $S_2(r) = S_1(r) - 1$.

- *release*: With the instruction

$$\ell \; : \; release(r) \; : \; \hat{\ell}$$

we associate all transitions $\langle S_1, S_2 \rangle$ where S_1 and S_2 are states in which control moves from ℓ to $\hat{\ell}$ as above, and $S_2(r) = S_2(r) + 1$.

In textbooks, usually queues of processes are introduced in order to explain by a concrete implementation what is a semaphore. The process executing a $P(S)$ instruction either decreases by 1 the value of semaphore variable S (in case this value is > 0), or else the process is suspended and added to a "waiting" queue from which a process can be "released" and resume its activity only when some other process executes a $V(S)$ operation. A $V(S)$ operation can either "release" a suspended process, which may then resume its activity, or else adds 1 to the value of the semaphore variable (when no processes are suspended). In case more than one process is suspended in the waiting queue the question which one gets released is left unspecified: not necessarily the first suspended is the first released.

Dijkstra avoids this pictorial description and his abstract specification uses no queue. The queue paradigm, however, expresses an important idea—namely that the P operation is executed "*as soon* as the resulting value would be nonnegative" (as Dijkstra says). This swiftness seems to me to be an important property of semaphores, and a main reason for their existence. It is completely absent from the cited Manna and Pnueli's book which does not require that the request operation is executed immediately—it may be delayed for an arbitrary long period. For example, suppose that a process tries to execute a request operation when the value of S is zero; it is not enabled to do so of course and it is suspended in the waiting queue. Afterwards another process executes a release operations, and then a third process makes a request. Now which process is going to get released, the first or the last to come (which made its request after the release)? It seems to me that Dijkstra intends that the first process, the one that is suspended, is released before the later process can even make its request. Yet in the Pnueli–Manna specification this is not necessarily the case.

I intend to prove that two seemingly different specifications of semaphores are in fact equivalent. The first is called the Textbook Specification in the following subsection. It is the traditional definition in which a queue of suspended processes is used. The second specification has a more abstract appearance, and I find that it is easier to use in correctness proofs. It is used here in the proofs for two protocols: the producer/consumer protocol in Section 4, and a critical section protocol in section 5.

3.1. The textbook specification. The semaphore instructions are viewed as parameterless external operations. $P(S)$ and $V(S)$ are a single symbol each. When the flowcharts of a program are drawn, each such instruction is translated into a one-arrow flowchart as shown in Figure 6.26. Transitions corresponding to the $P(S)$ or $V(S)$ instructions do not change any variable except the control positions. This corresponds to the intuition that, as far as the internal semantics

FIGURE 6.26. Flowcharts for $P(S)$ and $V(S)$ operations.

is concerned, the processes are not aware of the execution of these semaphore instructions, and of course they have no direct access to the semaphore's value. We have to specify the collection of system executions that describe correct operation of the semaphore.

The "textbook" specification uses queues of suspended processes to describe the behavior of semaphores. Two predicates P_S and V_S are defined on the events (S is the name of the semaphore). $P_S(e)$ says that e is a request (a call) to $P(S)$, and $V_S(e)$ says that event e is a release of the semaphore. We say that e is an S-event if $P_S(e)$ or $V_S(e)$ hold. S events can be terminating or non terminating. In fact, V_S events are always terminating, but P_S events may be non terminating. A non terminating P_S event corresponds to a suspended process that is never released. It is from the point of view of the executing process that a non terminating P_S event e is non terminating. It is the reason for the process inability to continue. For the system, however, the relevant part of this P_S event is very short. It consists of the $P(S)$ call followed by the subsequent suspension. This is certainly a terminating event. A suspended and never released $P(S)$ execution can thus be viewed as either a non terminating event that represents the infinite delay the process is incurring, or as a short call and suspension. From the point of view of the system the S events are so short that they are linearly ordered in time. To incorporate both views, we allow non-terminating $P(S)$ events, but we let $<$ be that linear ordering of the S events, which reflects the system view. Thus $e_1 < e_2$ may be the case for two P_S events even if e_1 is non terminating (whereas $e_1 \prec e_2$ always implies that e_1 is terminating).

The semaphore has a value S which is a natural number ≥ 0, and an associated queue Q_S of processes. If the set of processes is identified with $Id = \{1, \ldots, k_0\}$ then the value of the queue is a subset of Id, of "suspended" processes. Being a set rather than a sequence, Q_S should properly be called a *bag*. However, I use the traditional expression "queue of suspended processes" since in some applications one may want Q_S to be a queue with a FIFO discipline. The process of an event e is denoted $ProcId(e)$.

Each S event may change the value of the the semaphore and its queue. We therefore have four (unary) functions defined on the S events:

$$S, \ S', \ Q_S, \ Q'_S.$$

- $S(e)$ is the value of the semaphore when e begins, and $S'(e)$ is the resulting value.
- Similarly, $Q_S(e)$ is the value of the queue when e begins, and $Q'_S(e)$ is the resulting queue.

We shall use the following terminology in formulating the Textbook Specifications.

(1) If e is a P_S event (that is, $P_S(e)$ holds) and $S(e) > 0$, then e is called an *immediate* P_S event.
(2) If $P_S(e)$ but e is not immediate then e is said to be *suspended*. That is, e is suspended iff $S(e) = 0$.
(3) Suppose that e is a suspended P_S event and $k = ProcId(e)$. If s is a V_S event such that:
   (a) $e < s$ and $k \notin Q'_S(s)$.
   (b) For every S event e' with $e \le e' < s$,

$$k \in Q'_S(e').$$

Then we say that s *releases* e.
(4) If e is a suspended P_S event and no V_S event releases e, then e is said to be suspended forever. Otherwise, if e is suspended but also released, then we say that e is momentarily suspended.

The following six rules constitute the Textbook Specification:

T1 If e_1 is a P_S event then e_1 is terminating iff e_1 is immediate or else some V_S event $s > e_1$ releases e_1. In case s releases e_1, then $s \prec e_2$ for any event e_2 such that $e_1 \prec e_2$ and $ProcId(e_2) = ProcId(e_1)$ (e_2 is not necessarily an S event). In plain words, if a P_S event e_1 is suspended, then there are two cases. *Case one*: e_1 is suspended forever (never released), and then e_1 is non-terminating. In this case e_1 is the last event by its process. *Case two*: e_1 is released by some V_S event s. Then the process of e_1 can have no further events until its release, that is if e_2 is by the process of e_1 and $e_1 \prec e_2$ then $s \prec e_2$.

T2 The S events are partitioned into P_S and V_S events. All V_S events are terminating. $<$ is a linear ordering of the S events, and $s_1 < s_2$ if $s_1 \prec s_2$. If $e_1 < e_2$ are two S events such that e_2 is the successor of e_1 (no S event s with $e_1 < s < e_2$), then

$$S'(e_1) = S(e_2) \text{ and } Q'_S(e_1) = Q_S(e_2).$$

That is, the primed (resulting) value of an S event is the unprimed (initial) value of the following event. If s_0 is the first S event then $S(s_0) = 0$ and $Q_S(s_0) = \emptyset$. That is, the initial value of the semaphore is assumed 0, and the initial queue contains no suspended processes. In case when the initial value of the semaphore is $I > 0$, we simply assume I initial V_S events.

T3 If e is an immediate P_S event (that is, if $P_S(e)$ and $S(e) > 0$) then $S'(e) = S(e) - 1$ and $Q'_S(e) = Q_S(e)$. So this axiom says that an immediate P_S event decreases the semaphore value by 1 and does not change its queue.

T4 If e is a suspended P_S event (that is if $P_S(e)$ and $S(e) = 0$), then $S'(e) = 0$ and $Q'_S(e) = Q_S(e) \cup \{ProcId(e)\}$.

T5 If $V_S(e)$ and $Q_S(e) = \emptyset$, then $S'(e) = S(e) + 1$ and $Q'_S(e) = \emptyset$. In words: if a V_S event finds the queue of suspended processes empty, then it increases the semaphore by 1 (and the queue remains empty).

T6 If $V_S(e)$ and $Q_S(e) \neq \emptyset$, then for some process $k \in Q_S(e)$, $Q'_S(e) = Q_S(e) \setminus \{k\}$ and $S'(e) = S(e)$. In words: If a V_S execution finds that there are suspended processes, then it "releases" one of the suspended processes.

Let S be a system-execution that satisfies the Textbook Specification. For every terminating P_S event e, whether immediate or momentarily suspended, we shall associate a V_S event $s = \delta(e)$ which is the first V_S event responsible for the creation of the condition that made e immediate or released. We call δ the *permission* function. For a formal definition of δ we enumerate the S events in order:

$$s_0 < s_1 < \cdots .$$

Let $X_i = \{s_m \mid m \leq i\}$ be the initial segment of S events determined by s_i (X_i contains $i + 1$ elements).

We say that a P_S event e in X_i, with $k = ProcId(e)$, is *suspended in* X_i iff $k \in Q'_S(x)$ for every x in X_i such that $e \leq x$. It follows from T1 that X_i never contains two suspended events from the same process. It also follows that if e is not suspended in X_i, then it is not suspended in any X_j for $j > i$.

To define the permission function δ, we define $\delta|X_i = \delta \cap X_i \times X_i$ by induction on i. (A function is a set of ordered pairs, so that defining $\delta \cap X_i \times X_i$ means defining the relation $\delta(x) = y$ for x, y both in X_i.) Along this inductive definition we also prove the following properties P1 and P2 which are also used in the definition itself:

P1 $S'(s_i)$ equals the number of V_S events in X_i that are not in the range of $\delta|X_i$.

P2 The domain of $\delta|X_i$ is the set of all P_S events in X_i that are not suspended in X_i.

   $Q'_S(s_i)$ is the set of all $k \in Id$ such that $k = ProcId(s_j)$ for some $j \leq i$ such that s_j is suspended in X_i.

The definition of $\delta|X_i$ is now given in four cases. The domain of $\delta|X_i$ is the subset of P_S events in X_i that are not suspended there, and its range is a subset of V_S events in X_i.

D1 Suppose that s_i is an immediate P_S request (that is $S(s_i) > 0$). Then $i > 0$ since $S(s_0) = 0$, by T2. So by the inductive property P1 applied to s_{i-1}, $S'(s_{i-1}) = S(s_i)$ is the number (> 0) of V_S events in X_{i-1} that are not in the range of $\delta|X_{i-1}$. Define $\delta(s_i)$ to be the first such V_S event.

D2 Suppose that s_i is a suspended P_S request, that is that $S(s_i) = 0$. Then s_i is not in the domain of $\delta|X_i$.

D3 Suppose that s_i is a V_S event and $Q_S(s_i) = \emptyset$. Then s_i is not in the range of $\delta|X_i$.

D4 Suppose that s_i is a V_S event and $Q_S(s_i) \neq \emptyset$. By T6 there is some $k \in Q_S(s_i)$ such that $Q'_S(s_i) = Q_S(s_i) \setminus \{k\}$. As $k \in Q'_S(s_{i-1})$, P2 implies that there is a (unique) P_S event s_j with $j < i$ and $k = ProcId(s_j)$, such that s_j is suspended in X_{i-1} and hence is not in the domain of $\delta|X_{i-1}$. Well, define

$$s_i = \delta(s_j).$$

The proof of P1 and P2 involves several cases to check, but it is a routine which I think is best left to the careful reader. In addition to P1 and P2 the following properties of $\delta|X_i$ will also be needed.

P3 δ is one-to-one. The range of $\delta|X_i$ is an initial segment of V_S events in X_i.

P4   If $\delta|X_i$ is not defined on all P_S events in X_i, then its range is the set of all V_S events in X_i.

The term "initial segment" was used here. An initial segments of S events is a finite subset X of S that is downwards closed in S; that is

$$\forall x \in X \; \forall y \in S \; (y < x \rightarrow y \in X).$$

The empty set is an initial segment of S events and so is any set of the form

$$B(x) = \{e \mid e \in S \wedge e \leq x\}.$$

Except for the empty set any finite initial segment in S has the form $B(x)$ for some x (because every non-empty, linearly ordered and finite set of events has a maximum). Similarly the notion of an initial segment of V_S events can be defined as a downward-closed in V_S set of V_S events. An initial segment of P_S events is similarly defined.

The following lemma can be proved directly from the relevant definitions.

LEMMA 3.1. *If $\delta(e) = s_i$ and $e < s_i$ then s_i releases e. The other direction is also true: If s_i releases e then $\delta(e) = s_i$.*

P1–P4 imply for every S event s that $Q_S(s) \neq \emptyset \Rightarrow S(s) = 0$. Proof: If $Q_S(s_i) \neq \emptyset$, then $i > 0$ and P2 implies that some P_S event in X_{i-1} is suspended in X_{i-1}. Thus (again by P2) this suspended event is not in the domain of $\delta|X_{i-1}$. P4 now give that the range of $\delta|X_{i-1}$ is the set of all V_S events in X_{i-1}, and P1 now implies that $S'(s_{i-1}) = S(s_i) = 0$.

The properties of the permission function δ are summarized in the following lemma.

LEMMA 3.2. *Suppose that S satisfies the Textbook Specification, and δ is defined by D1–D4.*

(1) *δ is defined on all terminating P_S events, and its range is an initial segment of V_S events. δ is one-to-one.*

(2) *If $x \prec y$ are by the same process and x is a P_S event, then x is terminating and $\delta(x) \prec y$.*

(3) *If $e_1 < e_2$ are P_S events such that $\delta(e_2) < e_2$, then $e_1 \in dom(\delta)$ and $\delta(e_1) < \delta(e_2)$.*

(4) *If there is a non terminating P_S event, then the range of δ is the set of all V_S events.*

Proof. Most of the lemma appears in P1–P4 (locally, that is up to s_i). We prove here only item 3. So let $e_1 < e_2$ be two P_S events such that $\delta(e_2) < e_2$. Say $e_2 = s_i$. If e_1 is not in the domain of $\delta|X_{i-1}$, then by P4 the range of $\delta|X_{i-1}$ is the set of all V_S events in X_{i-1}. Thus $\delta(e_2) \in X_{i-1}$ is in this range, in contradiction to P3 which says that δ is one-to-one.

So $\delta(e_1)$ is defined in X_{i-1}. Since the range of $\delta|X_{i-1}$ is an initial segment of X_{i-1} (by P3), any V_S event (such as $\delta(e_2)$) not in this range is above $\delta(e_1)$.

3.2. Abstract specification of semaphores. The theory of semaphores introduced here is, as we shall see, equivalent to the textbook description, but it does not use numbers, and in fact semaphores have no values at all. The essence of this semaphore theory is that any process making a $P(S)$ request must, before continuing, find some releasing $V(S)$ event and that the function δ gives this event. If r and p are a request (namely a $P(S)$) event and a permission (namely a $V(S)$) event such that $p = \delta(r)$, then either $p < r$ or $r < p$. In the first case, $p < r$ indicates that permission was given in advance and the requesting process may continue immediately with its program. In the second case, $r < p$ indicates that the requesting process had to be suspended for a while, until permission p was given. In that case any event from the suspended process either precedes r or follows p. It is possible, of course, that a requesting process never obtains permission to continue, in which case it remains suspended for ever. In this case the request r is not in the domain of δ.

The signature for the Semaphore Theory contains two sorts: Events and Id. $Id = \{1, \ldots, k_0\}$ is the list of processes' names, which are taken as integers for simplicity. Two predicates P_S and V_S are defined on the events. We say that an event e is an S event iff $P_S(e) \vee V_S(e)$. (S is the semaphore's name.) A function $ProcId$ is defined on the events and gives for each event e the process $ProcId(e)$ which is said to have executed e. Then we have a partial function $\delta : Events \rightarrow Events$, and a binary relation $<$ on the events.

The following axioms form what we call the P/V theory.

The P/V Theory

PV1	For no S event can both $P_S(e)$ and $V_S(e)$ hold. All V_S events are terminating. $<$ is a linear ordering of the set of S events, extending $\prec$.
PV2	(1) The domain of δ is the set of terminating P_S events, and its range of values is an initial segment of V_S events.
	(2) δ is one-to-one. If $e_1 < e_2$ are P_S events and $\delta(e_2) < e_2$, then $e_1 \in dom(\delta)$ and $\delta(e_1) < \delta(e_2)$.
	(3) Either the domain of δ is the set of all P_S events or the range of δ is the set of all V_S events.
PV3	
	$$\forall x, y \; (x \prec y \wedge ProcId(x) = ProcId(y) \wedge P_S(x) \rightarrow$$ $$x \in dom(\delta) \wedge \delta(x) \prec y).$$
	In words: If x is a P_S operation, and y is any event in the same process that follows x, then $\delta(x)$ is defined and y follows $\delta(x)$.

FIGURE 6.27. The P/V axiom list.

I think that the intuitive meaning of these axioms can stand by itself, but to increase our belief that this theory fits our intuitive understanding of semaphores we show its equivalence with the textbook definition. We have already seen how to define δ in any system execution that satisfies the Textbook Specification, and Lemma 3.2 gives one direction of this equivalence.

For the other direction let S be a system execution that is a model of the P/V theory. The S events in S are serially ordered, and we enumerate them in order $s_0 < s_1 < \cdots$. (Since for any S events a and b $a \prec b$ implies $a < b$, the finiteness condition holds for $<$ and the sequence of s_i's is either finite or of the order type of N. Let $X_i = \{s_0, \ldots, s_i\}$ be the initial segment of the first $i + 1$ S events, and $\delta | X_i = \{\langle x, y \rangle \mid \delta(x) = y \wedge x, y \in X_i\}$.

We shall define two functions S' and Q'_S so that defining

$$\mathsf{S}(s_{i+1}) = \mathsf{S}'(s_i), \ Q_S(s_{i+1}) = Q'_S(s_{i+1}),$$

$$\mathsf{S}(s_0) = 0, \ Q_S(s_0) = \emptyset,$$

all axioms T1–T6 are satisfied.

DEFINITION 3.3. *Definition of* S' *and* Q'_S:

E1 $\mathsf{S}'(s_i)$ *is the number of* V_S *events in* X_i *that are not in the range of* $\delta | X_i$.

E2 *Let* $U_i \subseteq X_i$ *be the set of all* P_S *events* $u \in X_i$ *that are not in the domain of* $\delta | X_i$. *That is* $s_m \in U_i$ *iff* $m \leq i$, $P_S(s_m)$, *and either* $s_m \notin dom(\delta)$ *or* $s_i < \delta(s_m)$. *Observe that* U_i *cannot contain two events from the same process, because, by PV3, the first of any two such events would be in the domain of* $\delta | X_i$. *Define*

$$Q'_S(s_i) = \{ProcId(u) \mid u \in U_i\}.$$

LEMMA 3.4. *If* $k = ProcId(s_i)$ *and* $k \in Q'_S(s_i)$, *then* $s_i \in dom(\delta)$ *implies that* $s_i < \delta(s_i) = s_j$ *and* $k \notin Q'_S(s_j)$.

Proof. Assuming $k \in Q'_S(s_i)$ (for $k = ProcId(s_i)$) E2 implies that, for some $m \leq i$, s_m is a P_S event by process k such that

(10)   either $s_m \notin dom(\delta)$ or $s_i < \delta(s_m)$.

We claim that $m = i$. Otherwise, $m < i$ and then $s_m < s_i$ are both in process k, which implies by PV3 that $s_m \in dom(\delta)$ and $\delta(s_m) \prec s_i$. Which is a contradiction to (10), since $\delta(s_m) \prec s_i$ implies $\delta(s_m) < s_i$. Therefore $m = i$ and hence

either $s_i \notin dom(\delta)$ or $s_i < \delta(s_i)$.

In order to prove the lemma, assume $s_i \in dom(\delta)$ and let $\delta(s_i) = s_j$. We have just proved that $s_i < s_j$. It remains to prove that $k \notin Q'_S(s_j)$. By E2 this amounts to proving that every P_S event e by process k in X_j is in the domain of δ and $\delta(e) \in X_j$. But this is obvious from PV3. □

LEMMA 3.5. *If* s_i *is a* P_S *event and there is in* X_{i-1} *some* V_S *event that is not in the range of* $\delta | X_{i-1}$, *then* $s_i \in dom(\delta)$ *and* $\delta(s_i) < s_i$.

Proof. Assume the premises of the lemma and let v be a V_S event in X_{i-1} not in the range of $\delta|X_{i-1}$. We first prove that $s_i \in dom(\delta)$. For otherwise, PV2(3) implies that the range of δ is the set of all V_S events, and hence there is a P_S event x such that $v = \delta(x) \in X_{i-1}$. (So $x \notin X_{i-1}$.) Then $s_i < x$ and $\delta(x) < s_i$ contradicts PV2(2)!

Now we prove that $\delta(s_i) < s_i$. If this is not the case, then $s_i < \delta(s_i)$. Yet the range of δ is an initial segment of V_S events, and hence there is a P_S event x with $v = \delta(x)$. Since v is not in the range of $\delta|X_{i-1}$ and is not $\delta(s_i)$, $s_i < x$. Thus $\delta(x) < s_i < \delta(s_i)$ is in contradiction to PV2(2) (applied to $s_i < x$).

COROLLARY 3.6. *If s_i is an immediate P_S event, then $s_i \in dom(\delta)$ and $\delta(s_i) < s_i$.*

Proof. If $S(s_i) > 0$, then $i > 0$ and $S(s_i) = S'(s_{i-1})$. By E1 there is in X_{i-1} a V_S event that is not in the range of $\delta|X_{i-1}$. Now lemma 3.5 ends the proof. □

LEMMA 3.7. *If $\delta(s_i) < s_i$ then s_i is immediate.*

Proof. Put $\delta(s_i) = s_j$. Then $s_j \in X_{i-1}$, and as δ is one-to-one, s_j is not in the range of $\delta|X_{i-1}$. Thus $S'(s_{i-1}) > 0$ and as $S(s_i) = S'(s_{i-1})$, s_i is immediate. □

LEMMA 3.8. *If e is a P_S event such that $e < \delta(e)$, then $s = \delta(e)$ releases e and $s \prec e_2$ for any event e_2 such that $e \prec e_2$ and $ProcId(e_2) = ProcId(e)$.*

Proof. Let $e < \delta(e) = s_i$ be as in the lemma. "s_i releases e" means that $k \notin Q'_S(s_i)$, but for every S event s_ℓ with $e \leq s_\ell < s_i$, $k \in Q'_S(s_\ell)$, where $k = ProcId(e)$.

By E2, if $k \in Q'_S(s_i)$ then there is a P_S event $u < s_i$ with $ProcId(u) = k$ such that u is not in $dom(\delta|X_i)$.

- $e < u$ is impossible by PV3 since $u < \delta(e)$.
- $u = e$ is impossible since $e \in dom(\delta|X_i)$.
- $u < e$ is impossible by PV3.

So $k \notin Q'_S(s_i)$.

Let s_ℓ be any S event such that $e \leq s_\ell < s_i$. By E2, $k \in Q'_S(s_\ell)$ follows if we show that e is not in the domain of $\delta|X_\ell$. But this is obvious since $\delta(e) = s_i$. So s_i releases e.

Now if e_2 is any event by process k such that $e \prec e_2$, then PV3 applies and gives $\delta(e) \prec e_2$, which ends the proof of the lemma. □

LEMMA 3.9. *Let e be a P_S event and s_i a V_S event. If s_i releases e, then $\delta(e) = s_i$.*

Proof. Since s_i releases e, $e < s_i$. Let $k = ProcId(e)$. As $k \in Q_S(s_i)$ and $k \notin Q'_S(s_i)$, $k \in Q'_S(s_{i-1}) = Q_S(s_i)$. So E2 implies that there is a P_S event $u \in X_{i-1}$ with $k = ProcId(u)$ such that u is not in the domain of $\delta|X_{i-1}$ but u is in the domain of $\delta|X_i$. This leaves no alternative but

$$\delta(u) = s_i,$$

and our lemma follows if we show that $u = e$. We must rule our the two remaining possibilities $u \prec e$ and $e \prec u$.

- If $u \prec e$ then PV3 implies $\delta(u) \prec e$ which is a contradiction.
- If $e \prec u$ then PV3 implies again $\delta(e) \prec u$. But then $\delta(e)$ releases e (by the previous lemma), and it is not the case that s_i releases e, since clearly no two V_S events can release the same e. $\square$

Now we go over T1–T6 and prove that they hold in S.

T1 Suppose $P_S(e_1)$. e_1 is terminating iff $e_1 \in dom(\delta)$ (by PV2(1)). If e_1 is terminating, then either $\delta(e_1) \prec e_1$ (in which case e_1 is immediate) or $e_1 \prec \delta(e_1)$ and then $s = \delta(e_1)$ releases e_1.

T2 follows directly from PV1 and the definition of the functions S, S', Q_S, Q'_S.

T3 This follows from Lemma 3.5.

T4 Assume that s_i is some P_S event and $S(s_i) = 0$. We want to prove that $S'(s_i) = 0$ and

$$Q'_S(s_i) = Q_S(s_i) \cup \{ProcId(s_i)\}.$$

If $i = 0$ then there are no V_S events in $X_0 = \{s_0\}$ (since s_i is assumed to be a P_S event) and hence $S'(s_0) = 0$. Moreover, $U_0 = \{s_0\}$ (since s_0 cannot be in the domain of $\delta|X_0$ as $\delta(s_0)$ must be a V_S event). Hence $Q'_S(s_0) = \{ProcId(s_0)\}$.
Assume $i > 0$. Then $S(s_i) = S'(s_{i-1}) = 0$. Hence (by E1) all V_S events in X_{i-1} are in the range of $\delta|X_{i-1}$. As s_i is a P_S event, $S'(s_i) = 0$ as well. Moreover, s_i is not in the domain of $\delta|X_i$, or else δ would not be one-to-one. Thus $s_i \in U_i$ (by E2) and $ProcId(s_i) \in Q'_S(s_i)$ which gives the desired result.

T5 Suppose that $V_S(s_i)$ and $Q_S(s_i) = \emptyset$. We want to prove that $S'(s_i) = S(s_i) + 1$ and $Q'_S(s_i) = \emptyset$.
If $i = 0$ then s_0 is clearly not in the range of $\delta|X_0$, and hence $S'(s_0) = 1$ (as s_0 is a V_S event). Thus $S'(s_0) = S(s_0) + 1$ follows from $S(s_0) = 0$. As $U_0 = \emptyset$, $Q'_S(s_0) = \emptyset$.
Assume next that $i > 0$. Then $Q_S(s_i) = Q'_S(s_{i-1}) = \emptyset$. Thus (by E2) all P_S events in X_{i-1} are in the domain of δ. Hence the V_S event s_i cannot be in the range of $\delta|X_i$, and E1 now implies that $S'(s_i) = S(s_i) + 1$. As $U_i = U_{i-1} = \emptyset$ in this case, $Q'_S(s_i) = 0$.

T6 Assume that s_i is some V_S event such that $Q_S(s_i) \neq \emptyset$. We prove first that s_i is in the range of δ. Since $Q'_S(s_i) \neq \emptyset$, $i > 0$ and $Q_S(s_i) = Q'_S(s_{i-1})$. By E2, there is some P_S event u in X_{i-1} that is not in the domain of $\delta|X_{i-1}$. If u is not in the domain of δ, then the range of δ includes all V_S events (by PV2(3)) and hence s_i is in the range of δ. If u is in the domain of δ, then

$$\delta(u) \notin X_{i-1}$$

because otherwise u would be in the domain of $\delta|X_{i-1}$. Hence $s_i \leq \delta(u)$, and, since the range of δ is an initial segment of V_S events, s_i is in that range. Hence s_i is in the range of δ and $s_i = \delta(x)$ for some P_S event x.
We claim next that $x \in X_{i-1}$. If not, if $s_i < x$, then $u < x$ and $\delta(x) < x$ imply (by PV2(2)) that u is in the domain of δ and $\delta(u) < \delta(x)$. That is,

procedure $ENQUEUE(x : Item)$	procedure $DEQUEUE(\text{var } x : Item)$
(1) $P(Out)$;	(1) $P(In)$;
(2) $load(b, x)$;	(2) $unload(b, x)$;
(3) $V(In)$	(3) $V(Out)$

FIGURE 6.28. Dijkstra's protocol

$\delta(u) \in X_{i-1}$, which contradicts our choice of u as a P_S event in X_{i-1} not in the domain of $\delta | X_{i-1}$.

4. Load/unload with semaphores

Several solutions to the producer/consumer problem that use semaphores were devised by Dijkstra [13] and the simplest uses a single, safe buffer cell, and two semaphores Out and In. (See Figure 6.28.) The initial value of Out is 1 (and the initial value of In is 0). It turns out that these semaphores are binary: only values 0 and 1 are assumed, but this is something that has to be proved. Anyhow, in our approach semaphores have no values and an initial V_{Out} event replaces this initial value 1 assumption. The single buffer cell b is assumed to be safe.

Let S be any system execution for this protocol that is also a model for the P/V theory for both the Out and the In semaphores. So there are in fact two δ functions δ_{In} and δ_{Out}, defined on the terminating P_{In} and P_{Out} events respectively. For simplicity we use a single function symbol $\delta = \delta_{In} \cup \delta_{Out}$.

Let v_0 be the initial V_{Out} event (which exists by the assumption that the initial value of semaphore Out is 1). Recall that a P event (either P_{Out} or P_{In}) is terminating if and only if it is in the domain of δ.

Higher-level events are formed. $ENQUEUE$ and $DEQUEUE$ events represent executions of these procedures. *Producer* is the collection of $ENQUEUE$ events, and *Consumer* is the collection of $DEQUEUE$ events. There are two types of $ENQUEUE/DEQUEUE$ events: terminating and non-terminating, depending on the status of the P request. For example,

(1) A terminating $ENQUEUE$ event consists of three lower-level events in an execution of the $ENQUEUE$ protocol: a terminating execution of the $P(Out)$ instruction, a load event, and a $V(In)$ execution.

(2) A non-terminating $ENQUEUE$ is a singleton consisting of a non terminating $P(Out)$ event.

Similarly, a higher-level $DEQUEUE$ event is terminating if its P event (its first event) is terminating.

Each P_{Out} and V_{In} event is in some $ENQUEUE$ higher-level event, and each P_{In} and V_{Out} event is in some $DEQUEUE$ higher-level event (except for the initial V_{out} event denoted v_0 which is not in any higher level event).

For any higher-level terminating $ENQUEUE$ event E, define $\Delta(E)$ to be the following event (which is either v_0 or a terminating $DEQUEUE$ event): Let $e \in E$ be the (terminating) P_{Out} event. Look at $\delta(e)$, it is a V_{Out} event and hence it is either the initial v_0, or else $\delta(e)$ is in some $DEQUEUE$ event D. In the first case we define $\Delta(E) = v_0$ and in the second $\Delta(E) = D$.

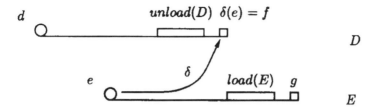

FIGURE 6.29. To the proof of Lemma 4.1.

In a similar manner, using the *In* semaphore, we define for every terminating *DEQUEUE* event D, a terminating *ENQUEUE* event $\Delta(D)$: Let $e \in D$ be the terminating execution of the $P(In)$ instruction. Then $\delta(e)$ is a V_{In} event, and we define $E = \Delta(D)$ where E is the unique higher-level *ENQUEUE* event containing $\delta(e)$.

Since δ is one-to-one, Δ is one-to-one. δ is not necessarily order preserving, but, restricted to the events in one process, δ *is* order preserving. Indeed, for any semaphore S, if $e_1 \prec e_2$ are P_S events by the same process, then $\delta(e_1) \prec e_2$ by PV3. If $e_2 < \delta(e_2)$ then $\delta(e_1) < \delta(e_2)$, and if $\delta(e_2) < e_2$ then $\delta(e_1) < \delta(e_2)$ by PV2(2). This implies the Δ is order preserving on the *ENQUEUE* and is order preserving on the *DEQUEUE* events.

For any terminating *ENQUEUE* (or *DEQUEUE*) event E (or D), let $load(E)$ ($unload(D)$) be the load (unload) event in E (or D). The meaning is clear: $load(X)$ or $unload(X)$ gives the buffer event in X.

LEMMA 4.1.   (1) *Let E be a terminating ENQUEUE event. Either $\Delta(E)$ is the initial V_{Out} event or else $D = \Delta(E)$ is a terminating DEQUEUE event and then:*
(a) *$unload(D) \prec load(E)$.*
(b) *$\Delta(D) \prec E$, and E is the successor of $\Delta(D)$ in Producer.*
(2) *If $E = \Delta(D)$ then E is a terminating ENQUEUE event and:*
(a) *$load(E) \prec unload(D)$.*
(b) *$\Delta(E) \prec D$, and D is the successor of $\Delta(E)$ in Consumer.*

Proof. To prove 1(a), suppose that $D = \Delta(E)$ is not the initial V_{Out} event. The proof follows the definition of Δ (and Figure 6.29). Let $e \in E$ be the P_{Out} event such that $\delta(e) \in D$. D is a *DEQUEUE* event. As $f = \delta(e)$ is a V_{Out} event, it is the last event in D and hence $unload(D) \prec f$. As $e \prec load(E)$, the third axiom PV3 implies that

$$\delta(e) \prec load(E).$$

Hence $unload(D) \prec load(E)$, which proves (a).

We now prove 1(b). As $D = \Delta(E)$ is terminating, $\Delta(D)$ is defined. $\Delta(D)$ is a terminating *ENQUEUE* event, and we claim that $\Delta(D) \prec E$. The additional claim that E is the immediate successor of $\Delta(D)$ (no other *ENQUEUE* event in between) will be an easy inductive consequence proved later.

Let $d \in D$ be the terminating request of $P(In)$, and let $\delta(d)$ be the corresponding $V(In)$ event. Since $d \prec f$ (d is the first and f the last event in D)

$$\delta(d) \prec f$$

follows from PV3. But we have already concluded that $f \prec load(E)$, and hence $\delta(d) \prec g$, where g is the last event in E—its V_{In} event. Now $\delta(d)$ is a $V(In)$ event, and it is in the $ENQUEUE$ event $\Delta(D)$. So as $g \in E$, $\delta(d) \prec g$ implies $\Delta(D) \prec E$.

Now the proof of the corresponding claims in (2) follows in a similar fashion, and we turn to the proof of the additional claims:

> For every terminating $DEQUEUE/ENQUEUE$ event X such that $\Delta(X)$ is again an $ENQUEUE/DEQUEUE$ event, X is the successor of the higher-level event $\Delta(\Delta(X))$.

Suppose that X is minimal to contradict this claim (minimal in the sense that if $X' \prec X$ then the claim holds for X'). We proved already that $\Delta(\Delta(X)) \prec X$, and so as the claim does not hold for X there is some higher-level event (in the same process of X and $\Delta(\Delta(X))$) with $\Delta(\Delta(X)) \prec Y \prec X$. We know that Δ is order-preserving, and hence

$$\Delta(Y) \prec \Delta(X).$$

$\Delta(Y)$ cannot be the initial event v_0 because it would then be preceded by $\Delta^3(X)$. So $\Delta(\Delta(Y))$ is defined, and order-preservation implies $\Delta(\Delta(Y)) \prec \Delta(\Delta(X))$. Hence

$$\Delta^2(Y) \prec \Delta^2(X) \prec Y,$$

and thus $Y \prec X$ is an earlier counterexample to the claim! This proves the lemma. □

We conclude in the lemma below that buffer b is conflict free in S and hence, by the assumed safeness, the value of any terminating $DEQUEUE$ event is the value of the last preceding $ENQUEUE$ event. So the kth terminating $DEQUEUE$ obtains the value of the kth $ENQUEUE$.

LEMMA 4.2. *The following holds in S.*

(1) *Between any two loads there is an unload event.*
(2) *Between any two unloads there is a load event.*
(3) *The first operation on the buffer is a load event.*

Proof. Any load event is of the form $load(E)$ for some terminating $ENQUEUE$ event E, and any unload event is of the form $unload(D)$ for some terminating $DEQUEUE$ event D.

Consider now two load events $load(E_1)$ and $load(E_2)$ where $E_1 \prec E_2$ are terminating $ENQUEUE$ events. Put $D = \Delta(E_2)$. Then D can either be the initial event V_0 or a $DEQUEUE$ event. The first possibility is ruled out since $\Delta(E_1) \prec \Delta(E_2)$ by order-preservation, implying that $\Delta(E_2)$ cannot be the initial event. Hence D is a terminating $DEQUEUE$ event and $\Delta(D)$ is defined. We know that $E_1 \preceq \Delta(D)$ (for otherwise $\Delta(D) \prec E_1 \prec E_2$ contradicts the previous lemma which says that E_2 is the successor of $\Delta(D)$). So

$$load(E_1) \preceq load(\Delta(D)) \prec unload(D),$$

and
$$unload(D) \prec load(E_2)$$
(by 1(b) and 2(b) of Lemma 4.1). This proves that an unload event exists in between the two given load events. A symmetric argument proves 2.

To prove that the first event on semaphore b is a load event (rather than an unload event) we prove that any unload event is preceded by a load. Let $unload(D)$ be an unload event. Then $\Delta(D)$ is defined, $\Delta(D) = E$ is an $ENQUEUE$ event, and $load(E) \prec unload(D)$. $\square$

The *liveness property* says that under the normal assumption of *Producer* and *Consumer* that repeatedly execute their $ENQUEUE/DEQUEUE$ operations, all $ENQUEUE$ and all $DEQUEUE$ are terminating. We prove by induction on k that the kth $ENQUEUE$ and the kth $DEQUEUE$ are terminating operations. To begin the induction, observe that since an initial V_{Out} event is assumed (namely v_0), the first P_{Out} event p_0 must be terminating. Indeed, if p_0 is non terminating then δ must be *onto* the V_{Out} events (by PV2(3)) and then $v_0 = \delta(x)$ for some P_{Out} event x. But since p_0 is the first P_{out} event, $p_0 \prec x$ which shows that p_0 is terminating. Hence the first $ENQUEUE$ operation is terminating.

We now prove that the first $DEQUEUE$ operation is terminating, and in fact we prove the general inductive $DEQUEUE$ step.

CLAIM 4.3. *If the kth $ENQUEUE$ operation is terminating, then the kth $DEQUEUE$ operation is terminating as well.*

Proof. Let E be the kth $ENQUEUE$ operation, supposed to be terminating, and let $w \in E$ be its V_{In} event. So w is the kth V_{In} event in the system execution. If the kth $DEQUEUE$ execution D is not terminating, then this is because its P_{In} evente is non terminating. But this implies that δ is onto the V_{In} events, and hence $w = \delta(x)$ for some P_{In} event x. Since δ is order-preserving on *Consumer* and as its domain and range are initial segments, x is the kth P_{In} event. Hence $x = e$ which is a contradiction.

CLAIM 4.4. *If the kth $DEQUEUE$ operation is terminating, then so is the $(k+1)$th $ENQUEUE$ operation.*

Proof. Let D be the kth $DEQUEUE$ operation. Let v_k be the V_{Out} event in D. Then v_k is the $(k+1)$th V_{Out} event in the system execution because of the initial event v_0. If the $(k+1)$th $ENQUEUE$ operation E is not terminating, then it consists of a non terminating P_{Out} event p. This implies that δ is onto the V_{Out} events and so $v_k = \delta(x)$ for some P_{out} event x. This x must be the $(k+1)$th P_{Out} event and hence $x \in E$, and $x = p$, which is a contradiction.

5. A Multiple Process Mutual Exclusion Protocol

Obtaining mutual exclusion with the aid of a semaphore is very easy, even for a multiple process environment. Suppose k_0 processes: $P_1, \ldots, P_{k_0}$ executing concurrently the protocol of Figure 6.30.

A_i and B_i are assumed to be terminating operations, but CS_i (the critical section events) may be non terminating. S is some semaphore initially set to 1 (that is, there is an initial V_S event). Let δ be the permission function of

procedure P_i	procedure $Round_i$
begin	**begin**
repeat_forever	(1) A_i;
$Round_i$	(2) $P(S)$;
end	(3) CS_i;
	(4) $V(S)$;
	(5) B_i
	end

FIGURE 6.30. Multiple Process Critical Section Protocol, and declaration of *Round*.

S, and $<$ be the linear ordering on the S events (satisfying the P/V theory). Let $\mathcal{S}$ be a system execution resulting from this protocol. So each process P_i is executing "rounds" consisting each of an external activity A_i, followed by a $P(S)$ call to semaphore S, which (when permission is granted) leads into a critical section execution, and finally a $V(S)$ release and some other business B_i that end the round. A round is *alive* iff its P_S event is terminating. An alive round is terminating if its CS event is terminating, because all other instruction executions are assumed to be terminating. Our aim here is to prove the mutual exclusion property for the CS_i events, and the deadlock freedom for the protocol. That means that, if each critical section event is terminating, then there are infinitely many such events. It is not necessarily true that every P_S event is terminating, but we shall see that there are infinitely many such events and hence infinitely many alive rounds unless there is a non terminating CS event. So the situation in which every process P_i contains a non terminating P_S event is impossible.

For any terminating P_S event e, if the CS event in the round of e is terminating, then there is a later V_S event $s = \sigma(e)$ in the same round. Indeed, since e is followed by a terminating CS event, the process executes line (4) and thence the V_S event which is denoted $\sigma(e)$. If the CS event in the round of e is non terminating, then $\sigma(e)$ is undefined. Clearly σ is one-to-one.

LEMMA 5.1. *If e is a terminating P_S event, then either $\delta(e)$ is the last V_S event and in this case the CS event in the round of e is non terminating, or else $\sigma(e)$ is defined, $\delta(e) \prec \sigma(e)$, and $\sigma(e)$ is the successor of $\delta(e)$ in the $<$ ordering on the V_S events. (See the P/V Theory on Figure 6.27.)*

Proof. Let e be a terminating P_S event that contradicts the lemma, and assume that $\delta(e)$ is minimal. That is, if e' is a terminating P_S event such that $\delta(e') < \delta(e)$, then e' satisfies the lemma.

Assume first that the CS event in the round of e is non terminating. Since e contradicts the lemma, $\delta(e)$ is not the last V_S event (in $<$). Let s be the first V_S event such that $\delta(e) < s$ (namely the successor of $\delta(e)$ in the $<$ sequence of V_S events). Then s is not the initial V_S event, and thus s is in some round (but not in the round of e since this non terminating round contains no V_S event). Let e_1 be the P_S event in the round of s. Then $s = \sigma(e_1)$ and $\delta(e_1) \prec s$ by PV3 (since e_1 and s are in the same round and thence by the same process).

Since δ is one-to-one, $\delta(e) \neq \delta(e_1)$. $\delta(e) < \delta(e_1)$ would contradict the minimality of s, and hence $\delta(e_1) < \delta(e)$. But e_1 itself is also a counterexample to the lemma (as $\delta(e_1) < \delta(e) < s = \sigma(e_1)$ shows that $\sigma(e_1)$ is not the $V_S <$ successor of $\delta(e_1)$). This contradicts the minimality of $\delta(e)$.

Assume next that the CS event in the round of e is terminating, and hence that $\sigma(e)$ is defined. As e and $\sigma(e)$ are by the same process, $\delta(e) \prec \sigma(e)$ by PV3 (and so $\delta(e) < \sigma(e)$). Since we assume that e is a counterexample, $\sigma(e)$ is not the successor V_S event of $\delta(e)$, and hence there is a V_S event s such that

$$\delta(e) < s < \sigma(e).$$

We take s to be the minimal such V_S event. Let e_0 be the unique P_S event such that $\sigma(e_0) = s$. (As s is a V_S event and it is not the initial V_S event, it is in some round, and e_0 is simply the terminating P_S event in that round.) Then $\delta(e_0) \prec s$ by PV3. $\delta(e) < \delta(e_0) < s$ is impossible, since s is the successor V_S event of $\delta(e)$. Hence $\delta(e_0) \leq \delta(e)$. This leads to a contradiction as follows.

- $\delta(e_0) = \delta(e)$ is impossible since it would imply $e = e_0$, which is not the case as $\sigma(e_0) = s < \sigma(e)$.
- $\delta(e_0) < \delta(e)$ is also impossible because $\delta(e_0) < \delta(e) < s = \sigma(e_0)$ would show that $\delta(e_0)$ is an earlier counterexample.

COROLLARY 5.2. *The mutual exclusion property holds for the critical section events. That is if $C_1 \neq C_2$ are CS events, then $C_1 \prec C_2$ or $C_2 \prec C_1$.*

Proof. Let $C_1 \neq C_2$ be two CS events, and let e_1, e_2 be the terminating P_S events in the rounds of C_1 and C_2 respectively. Then

$$e_i \prec C_i \text{ for } i = 1, 2.$$

Clearly $e_1 \neq e_2$ and hence $\delta(e_1) \neq \delta(e_2)$. So $\delta(e_1) < \delta(e_2)$ or $\delta(e_2) < \delta(e_1)$. We shall see that this determines $C_1 \prec C_2$ or $C_2 \prec C_1$ respectively.

Assume that $\delta(e_1) < \delta(e_2)$. So $\delta(e_1)$ is not the last V_S event. By the above lemma this implies that C_1 is terminating and $\sigma(e_1)$ is the successor of $\delta(e_1)$. Hence $\sigma(e_1) \leq \delta(e_2)$. So we have

$$C_1 \prec \sigma(e_1) \leq \delta(e_2) \prec C_2.$$

($\delta(e_2) \prec C_2$ by PV3.) But now $\sigma(e_1) \leq \delta(e_2)$ implies $\sigma(e_1) \vdash \delta(e_2)$, and the Russell–Wiener property implies that $C_1 \prec C_2$. $\square$

The liveness property is proved as follows. Assume that all CS events are terminating, but there is a finite number of terminating rounds, and we shall derive a contradiction. Since all CS events are terminating, each alive round is terminating. The number of terminating P_S events is finite, because otherwise there are infinitely many alive and terminating rounds. Thus each process contains a non terminating P_S event, and by PV2(3), δ is onto the V_S events, and hence there is a finite number of V_S events. Since there is an initial V_S event, this set is non empty. Let v be the $<$ last V_S event, and let x be that P_S event such that $\delta(x) = v$. But then $\delta(x) \prec \sigma(x)$ follows from PV3 and contradicts the maximality of v! $\square$

7

Circular buffers

We continue our discussion of the Producer/Consumer problem from the last chapter. Two versions of the circular buffer protocol are given: the first uses unbounded sequence numbers for coordination, and the second only $N + 1$ numbers to control access to an array of N buffer cells.

The Producer/Consumer problem was specified in the previous chapter. A generic protocol was written there which left operations $ENQUEUE/DEQUEUE$ unspecified, and a solution which uses semaphores was described. Here we implement $ENQUEUE/DEQUEUE$ operations with a protocol that uses registers rather than semaphores to control the buffer's accesses. These registers carry unbounded numbers in the first solution, and only $N + 1$ numbers in the second solution, where N is the number of buffer cells. The protocols described here are quite standard and similar protocols can be found in any textbook on operating systems (for example, Silberschatz et al. [28]).

1. Unbounded sequence numbers

The protocol is in Figure 7.31. There are two serial processes *Consumer* and *Producer* which use two single-writer single-reader serial registers: *In* (from *Producer* to *Consumer*) and *Out* (from *Consumer* to *Producer*); these registers carry natural numbers, and initially have value 0. Register *In* records the number of items produced and *Out* the number consumed. We use read/write commands with the register's name as parameter. For example, $Write(In, in)$ is the command to write into register *In* the value of variable *in*.

buf is an N-place array of "cells" $buf[0], \ldots, buf[N-1]$, where $N \geq 1$, which carry data values of type *Item*. The cells are loaded by *Producer* and unloaded by *Consumer*. The instruction to load $buf[i]$ with value x is

$$load(buf[i], x)$$

and the instruction to unload $buf[i]$ into variable x is

$$unload(buf[i], x).$$

procedure $ENQUEUE$ $(x : Item)$	procedure $DEQUEUE$ $(var\ x : Item)$
var $in,\ out$: N; p: $0..N-1$;	var $in,\ out$: N; p: $0..N-1$;

(1) while $in - out = N$ do
     $Read(Out, out)$;
	(1) while $in - out = 0$ do
	     $Read(In, in)$;

(2) $load(buf[p], x)$;	(2) $unload(buf[p], x)$;
(3) $p := p + 1$ (mod N);	(3) $p := p + 1$ (mod N);
(4) $in := in + 1$;	(4) $out := out + 1$;
(5) $Write(In, in)$	(5) $Write(Out, out)$

FIGURE 7.31. Unbounded $ENQUEUE/DEQUEUE$ protocol. The initial value of all variables is 0.

We define the *value* of the load/unload event to be the value of the item x, loaded or unloaded, and the function *Value* is used in any system execution to give this value of load/unload events. We assume safeness for each $buf[i]$ (see Chapter 6 Section 2).

The local variables, in, and out carry natural numbers, and are initially 0. The local variable p has values in $0..N - 1$, and is initially 0. (Any initial value works, but starting with 0 simplifies some of the notations used in the proof.) (Locality means of course that these are distinct variables when they belong to different procedures, even if apparently they have the same name. So we deviate somewhat from the strict syntax of chapter 3, and follow the common practice of using the same name for variables that have the same function.) It is exactly the role of variable declarations to allow such practice, but since our simple languages have no variable declarations we must rely here on the reader's understanding.)

Observe that in executing $ENQUEUE$ the *Producer* does not always have to read Out: it may rely on condition $in - out \neq N$ to access the buffer immediately. Similarly, an execution of $DEQUEUE$ does not always necessitate reading register In. This is an advantage in case variables are accessed more rapidly than registers.

Observe also that we use here two types of variables: in and out vary over the set of natural numbers N, while p varies through the indices of the buffer array. It is not difficult to prove that variable p is dispensable and $\lceil in \rceil_N$ (for $ENQUEUE$) or $\lceil out \rceil_N$ (for $DEQUEUE$) could be used instead of p (where $\lceil k \rceil_N$ is the remainder of k (mod N)). We use variable p for clarity and since it will have a distinctive role in the bounded protocol of Section 2.

Both the $ENQUEUE$ and the $DEQUEUE$ procedures can be decomposed into three parts:

(1) the *wait* part, consisting of line 1, which may be nonterminating,

(2) a load/unload event which is always terminating, and

(3) the remainder consisting of lines (3) (4) (5).

Let $\mathcal{H}$ be a system execution resulting from an application of the protocol to the *Producer/Consumer* program (Figure 1 of Chapter 6). H is the set of events, and $\prec$ is the precedence ordering on H. The events in H are the lower-level read/write events on the registers (assumed to be serial), the load/unload events on the buffer cells, as well as the higher-level operation executions obtained by grouping lower level events as explained in Chapter 3. $X \in H$ in an *ENQUEUE* execution if X is a set of lower-level events that constitutes an execution of the *ENQUEUE* protocol (for some parameter x of type **Item**). Similarly *DEQUEUE* executions are introduced. As in Chapter 5, Section 3, we make a difference between executions and events: As we shall define below, higher-level *ENQUEUE* events contain only those lower-level events that matter (in a sense) and similarly higher-level *DEQUEUE* events shall be defined.

Registers *In* and *Out* are assumed to be serial. Let ω be the function that assigns to every read r of the registers *In* or *Out* the corresponding write $\omega(r)$ which is the $\prec$ rightmost such that $\omega(r) \prec r$. $Value(r) = Value(\omega(r))$. We assume that $\mathcal{H}$ contains initial writes w_{In} and w_{Out} of value 0 onto registers *In* and *Out*, so that $\omega(r)$ is always defined.

An execution of the *ENQUEUE/DEQUEUE* procedure is called *successful* if it passed the while loop of line 1 and arrived to line 2 (and thence to the end of the protocol as load/unload events are always terminating); it is *unsuccessful* otherwise. Thus an execution is successful either because the while loop is not activated or else because it terminates. (There can be at most one unsuccessful *ENQUEUE*, and at most one unsuccessful *DEQUEUE* event, because these are non terminating events.)

Let $X \in H$ be a successful *DEQUEUE* or *ENQUEUE* operation execution; it is possible that the condition in the while loop of X does not hold, and no read is necessary for its completion. We say that X is *immediate* in this case. If, however, the body of the while loop of X is executed, then we call the last read in the loop of register *In* or *Out* the *successful read*. The successful read is thus the read in line 1 which ends the while loop of the successful event.

Now if X is a successful, non-immediate *DEQUEUE* or *ENQUEUE* operation execution and the lower-level events in X preceding the successful read are omitted, then the resulting set of events is called a *DEQUEUE* or *ENQUEUE* (successful) event. Thus the first lower-level event in a *DEQUEUE* (or *ENQUEUE*) event is its successful read. Omitting these non-successful reads is only for convenience: some of the statements in the proof appear more natural with these reduced higher-level events. It is by no means an assumption that we are making or a limitation of the results proved.

If X is an immediate execution, then it contains no reads of the register, and then X itself is the *ENQUEUE/DEQUEUE* event. In case X is not successful, X contains an infinite set of read events which never satisfy the loop's termination condition. The corresponding event is again X itself. thus an *ENQUEUE/ DEQUEUE* event is different from the corresponding execution just in case some non-successful reads are omitted.

The higher-level *ENQUEUE* and *DEQUEUE* events in $\mathcal{H}$ are enumerated in increasing $\prec$ order as $\langle E_i : i \in I_0 \rangle$ (in *Producer*), and $\langle D_j : j \in J_0 \rangle$ (in *Consumer*), where I_0 (and J_0) can either be a finite initial segment or the complete set of the natural numbers. So E_0 is the first *ENQUEUE* event and D_0 is the first *DEQUEUE* event. E_1 and D_1 are the second *ENQUEUE* and *DEQUEUE* events, etc. If we want to refer to the execution, rather than the event, then E_i^+ and D_j^+ are used for the full *ENQUEUE* and *DEQUEUE* executions.

If I_0 is finite and $i_0 = \max(I_0)$, then E_{i_0} is not successful (hence non-terminating), but E_i is always successful if $i + 1 \in I_0$. Similarly for J_0; only the last event (if there is one) is non-terminating.

DEFINITION 1.1 (*load* AND *unload*). *The functions load and unload are defined on the successful higher-level ENQUEUE/DEQUEUE events. If X is a successful ENQUEUE event, we denote with load(X) the load event in X (corresponding to line 2). Similarly, unload(X) is defined for a successful DEQUEUE event X. Since each successful ENQUEUE/DEQUEUE events contains a unique load/unload event, this function is well-defined. The values of these load/unload events are also considered as the values of the higher-level ENQUEUE/DEQUEUE events containing them, and we write Value(X) for the value of load(X) or unload(X).*

We want to prove the following two properties.

Safety Property: For every n for which D_n exists, E_n exists as well, $E_n \prec D_n$ and both have the same value.

Liveness property : The *ENQUEUE/DEQUEUE* events are all successful. Hence $I_0 = J_0 = \mathbf{N}$.

There are additional (and finer) properties that one may want to formulate: For example, if $n \geq N$ and E_n is successful, then the *Producer* is ensured that for each $i \leq n - N$ the load made in E_i has been unloaded (by D_i).

(Already here one can see the advantage of throwing away all non-successful reads in defining the higher-level events. It is not necessarily true that the nth *ENQUEUE* operation execution precedes the n *DEQUEUE* operation execution, even if these are successful operations. The problem is with those non-successful reads in the *DEQUEUE* operation which may spoil this precedence.)

Notice that when line 1 of the *ENQUEUE* protocol is executed for the very first time, $in = 0$ and $out = 0$ are the initial values, and hence condition

$$in - out = 0$$

holds (in E_0^+), and as $0 < N$ the body of the **while** loop is not activated (that is, register Out is not read). Similarly for all $n < N$, condition

$$in - out = n$$

holds at line 1 of E_n^+, which does not activate the loop's body. Only for E_N^+ (the $N + 1$-th execution), $in = N$ does hold and the *ENQUEUE* protocol requires reading register Out, to make sure that there are any free buffer cells. Thus E_N is not immediate. In contrast, *DEQUEUE* requires reading register In at the first execution, since its initial conditions are $in = 0$ and $out = 0$, so that already D_0 is not immediate.

If (against our basic assumption) the *Producer* is not active at all, then no *DEQUEUE* event may be successful. However we assumed that *Producer* is active forever in executing its **repeat_forever** loop of producing and enqueueing. Moreover, each produce_an_item event is terminating. Thus non zero values are written on In, and the first execution of the *DEQUEUE* procedure must at some stage read register In and obtain a value $\neq 0$. This successful read enables the execution of the rest of the protocol, and hence the first *DEQUEUE*, $D_0 i$, is successful. We will show (in subsection 1.3) that, as the processes are active forever, every procedure's execution are is successful: the **while** loop is always terminating. However, it is convenient at this stage of the proof to speak about possible unsuccessful executions.

We have assumed that the buffer cells *buf[i]* are safe. This safeness assumption is the "hardware" assumption needed to ensure the correctness of the protocol.

We use the notation $\lceil i \rceil_N$ to denote that number $0 \leq \lceil i \rceil_N < N$ such that $i = \lceil i \rceil_N$ (mod N), for every i.

1.1. The function *activate*. The following definitions refer to a system execution $\mathcal{H}$ of our protocol.

DEFINITION 1.2 ($in(X)$, $out(X)$, $In(X)$, $Out(X)$). *For any successful operation execution $X \in H$, define $in(X)$ and $out(X)$ as the value of these variables exactly at the end of executing line 1 in X. For any successful ENQUEUE X, $In(X)$ is the value written by X on register In (in executing line 5). Similarly, for any successful DEQUEUE X, $Out(X)$ is the value written by X onto register Out.*

We repeat the definition to make sure that it is clear. If E is a successful *ENQUEUE* event, then the value of variable *in* at the beginning of E and at the end of execution of the **while** loop (line 1) is the same. However, the value of *out* may change, because it may be determined by a read of *In*. $out(E)$ is defined expressly as the value at the end of executing line 1. Similarly if D is a *DEQUEUE* event, then $out(D)$ could be defined as either the value at the beginning of execution of line 1 or at its end, it does not matter since *out* is changed only at line 4, but the definition of $in(D)$ requires the value at the end of the **while** loop.

LEMMA 1.3. *The following holds in $\mathcal{H}$.*

(1) $in(E_0) = 0$; $out(D_0) = 0$. *For any E_i*

$$in(E_i) = i, \text{ and } In(E_i) = i + 1.$$

For any D_j,

$$out(D_j) = j \text{ and } Out(D_j) = j + 1.$$

(2) *If D is a successful DEQUEUE, then*

$$in(D) \neq out(D).$$

If E is a successful ENQUEUE, then

$$in(E) - out(E) \neq N.$$

The **proof** of (1) is by induction on i, and the proof of (2) is by observing that a successful execution of a loop "**while** τ **do** s" leaves τ in a state where it is false. Hence, if D is a successful *DEQUEUE*, then, at the end of executing line

1, condition "$in - out = 0$" is false. Namely $in \neq out$ is true. As the values of these variables at the end of line 1 are denoted $in(D)$ and $out(D)$, we get

$$in(D) \neq out(D).$$

As trivial as it sounds, a formal proof of this lemma requires the apparatus of Chapter 3 (or some equivalent framework).

DEFINITION 1.4 ($activate(X)$). *The function $activate(X)$ is inductively defined on the successful DEQUEUE events in H, and on the successful ENQUEUE events E_n with indices $n \geq N$, as follows. Let X denote any such event. The definition splits into two cases depending on whether X contains or not a successful read.*

- *If X contains a successful read r, let $\omega(r)$ be the corresponding write. In case $\omega(r)$ is the initial write on its register, let*

$$activate(X) = \omega(r).$$

Otherwise, $\omega(r)$ is a write in some (uniquely determined) higher-level event Y, and we define

$$activate(X) = Y.$$

In particular, as E_N and D_0 are not immediate, the definition of $activate(E_N)$ and of $activate(D_0)$ fall under this case. Namely, if E_N is successful (and if D_0 is successful) then it must contain a successful read r. We argued that E_N and D_0 are not immediate (in the paragraph starting with Remark on page 135) and then $activate(E_N)$ and $activate(D_0)$ are determined by $\omega(r)$.

- *If X is immediate, and thus contains no read of In or Out, then define $activate(X) = activate(Y)$ where Y is the predecessor of X in its process. That is, either $X = E_i$ for $i > N$ or $X = D_i$ for $i > 0$, and we define accordingly*

$$activate(E_i) = activate(E_{i-1}), \text{ or } activate(D_i) = activate(D_{i-1}).$$

LEMMA 1.5.   (1) *For every X in the domain of activate:*

$$activate(X) \prec X.$$

(2) *If X is a successful ENQUEUE in the domain of activate, then $activate(X)$ is a successful DEQUEUE. Similarly, if X is a successful DEQUEUE, then $activate(X)$ is a successful ENQUEUE.*

(3) *activate is monotonic: If $N \leq i \leq i'$ then*

$$activate(E_i) \preceq activate(E_{i'}),$$

and if $0 \leq i \leq i'$ then

$$activate(D_i) \preceq activate(D_{i'}).$$

(4) *Let D and E be a DEQUEUE and ENQUEUE successful events:*

$$If\ E = activate(D),\ then\ in(D) = In(E).$$

$$If\ D = activate(E),\ then\ out(E) = Out(D).$$

Proof. The higher-level events are enumerated, and each of the items of this lemma can be proved by induction on the index of the event in this enumeration. The first item, $activate(X) \prec X$, is proved by induction on i for $X = E_i$ or $X = D_i$. If X contains a successful read and r is that read, then $activate(X) \prec r$ follows from the seriality of the registers, the fact that the write on the register is the last event in every ENQUEUE/DEQUEUE operation execution. Then $activate(X) \prec X$ follows from the fact that the successful read r is the first event in X. (It is for this lemma that the unsuccessful reads were omitted from the events.)

In case X is immediate and contains no successful read, then $activate(X) = activate(Y)$ where Y is the predecessor of X in its process. Then by the inductive assumption $activate(Y) \prec Y$ holds, from which $activate(X) \prec X$ follows.

To prove the second item assume that $activate(X)$ is defined where X is an ENQUEUE event for example. A priori, $activate(X)$ may be an initialization write event, and not necessarily a DEQUEUE event. So there is something to prove here and we do it by induction: For E_N, if E_N is successful and r_0 is its successful read, then $\omega(r_0)$ cannot be the initial write on Out (as this write is of value 0 and condition $in(E_N) - out(E_N) \neq N$ must hold). Thus $\omega(r_0)$ is some later write on Out which must be in some DEQUEUE event. Now for $n > N$ there are two cases to consider for $activate(E_n)$. If E_n contains a successful read r, then $E_N \prec E_n$ implies that $\omega(r_0) \preceq \omega(r)$, and hence $\omega(r)$ is not the initial write. If E_n contains no successful read, then $activate(E_n) = activate(E_{n-1})$ by definition, and induction gives the result.

The third item is proved by induction on i' as follows. Assume that $N \leq i \leq i'$, and we want to prove that $activate(E_i) \preceq activate(E_{i'})$. If $i = i'$ then the lemma is trivially true. Else, $i < i'$ and the inductive assumption can be applied to $i \leq i' - 1$ to deduce that

$$activate(E_i) \preceq activate(E_{i'-1}).$$

Now there are two cases corresponding to the two cases in the definition of $activate(E_{i'})$. The simpler is when $E_{i'}$ s immediate and then $activate(E_{i'-1}) = activate(E_{i'}')$ implies the lemma. In the second case $E_{i'}$ contains a successful read r. The first item shows that $D = activate(E_{i'-1}) \prec E_{i'-1}$, and hence that

$$D \prec E_{i'}.$$

(Because $E_{i'-1} \prec E_{i'}$.) But this implies $D \preceq activate(E_{i'})$ (because $\omega(r) \prec D$ is impossible) and hence the result.

The fourth item is proved by induction. For example, assume that $E = activate(D_i)$, and we will prove that $in(D_i) = In(E)$ by induction on i. In case D_i contains a successful read r then $E = activate(D_i)$ is defined by requiring that E contains the write $\omega(r)$. It follows from our assumptions on ω that the value of r is the same as the value of $\omega(r)$. But the value of $\omega(r)$ is $In(E)$

the value of register In in E, and the value of r is $in(D_i)$ by definition. Hence $In(E) = in(D)$.

In case D_i contains no reads of In, then $activate(D_i) = activate(D_{i-1})$ and the inductive assumption yields the lemma since $in(D_{i-1}) = in(D_i)$ in this case. $\square$

The higher-level $load/unload$ events, and the functions with these same names were already defined in page 135 suppose that $E_m = activate(D_i)$. Then $E_m \prec D_i$ was established in Lemma 1.5. As $load(E_m)$ is an event in E_m, and $unload(D_i)$ an event in D_i, $load(E_m) \prec unload(D_i)$ follows. We thus have the following:

LEMMA 1.6. (1) If $E_m = activate(D_i)$, then $load(E_m) \prec unload(D_i)$.
(2) If $D_n = activate(E_k)$, then $unload(D_n) \prec load(E_k)$.

The following two lemmas are the main lemmas to prove. They will be used to show that at no time does the number of $DEQUEUE$ operations exceeds the number of $ENQUEUE$ operations, and at no time does the number of $ENQUEUE$ operations exceeds by more than N the number of $DEQUEUE$ operations. We shall need later the following fact which the reader can check: that only properties established in lemmas 1.3 and 1.5 are used in the proofs below.

LEMMA 1.7 (MAIN LEMMA). For every $i \geq N$, if E_i is successful and $D_j = activate(E_i)$, then $i \leq j + N$.

Proof. By induction on i: If $i = N$, then the lemma is obvious since $j \geq 0$. Suppose now that $D_j = activate(E_i)$ and that the lemma holds for all values below i. Assume that

$$j + N < i$$

and we will derive a contradiction. So $j + N < i$ is assumed, and the lemma holds for $j + N$ by the induction hypothesis. Say $D_{j_0} = activate(E_{j+N})$ and then $j + N \leq j_0 + N$ follows from the lemma, or equivalently

$$j \leq j_0.$$

On the other hand, since $j + N < i$,

$$activate(E_{j+N}) = D_{j_0} \preceq D_j = activate(E_i)$$

by monotonicity of $activate$. Hence $j_0 \leq j$, (or else $j < j_0$ would imply $D_j \prec D_{j_0}$). This gives $j = j_0$ because $j \leq j_0$ was established above. That is,

$$activate(E_{j+N}) = D_{j_0} = D_j = activate(E_i).$$

Hence

$$D_j = activate(E_{j+N+1})$$

by monotonicity again (a sandwich lemma applied to $E_{j+N} \prec E_{j+N+1} \preceq E_i$). This implies that $out(E_{j+N+1}) = Out(D_j)$ by Lemma 1.5(4), and hence Lemma 1.3(1) implies

$$out(E_{j+N+1}) = Out(D_j) = j + 1.$$

However this is not possible because $in(E_{j+N+1}) = j + N + 1$, and so condition $in - out = N$ holds for E_{j+N+1} which is therefore not successful! (See Lemma 1.3(2).) $\square$

In a similar way the following lemma can be proved.

LEMMA 1.8. *For any $j \geq 0$ if $E_i = activate(D_j)$ then $j \leq i$.*

Exercise: Prove the lemma by induction on j (or see Lemma 2.3 below).

1.2. Safety. The safety property says that for every n for which D_n exists, E_n exists as well, $load(E_n) \prec unload(D_n)$ and both have the same value. Fixing some arbitrary $0 \leq p < N$, it suffices to prove this claim for all n's such that $\lceil n \rceil_N = p$. We are going to prove that the load/unload operations on $buf[p]$ are conflict free. Since the safeness of $buf[p]$ is assumed, this implies that every unload event obtains the value of the last preceding load event on $buf[p]$. But since the only load/unload operations on $buf[p]$ are those included in E_n and D_n for $n = p \pmod{N}$, it suffices to prove that:

(11) $j \in J_0$ and D_j is successful $\Rightarrow j \in I_0$, E_j is successful, and $E_j \prec D_j$,

(12) $i + N \in I_0$ and E_{i+N} is successful $\Rightarrow i \in J_0$ and $D_i \prec E_{i+1}$

These equations show that $buf[p]$ is conflict free:

$$load(E_p) \prec unload(D_p) \prec load(E_{p+N}) \dots$$
$$\dots load(E_i) \prec unload(D_i) \prec load(E_{i+N}) \dots .$$

For example, to show that between any two load events of $buf[p]$ there is an unload event, we may take two successive loads which must be in some E_i and E_{i+N} where $i = p \pmod{N}$. But $E_i \prec D_i \prec E_{i+N}$ follows from equation (11) and (12) above. So, since each $buf[p]$ is conflict free, and as the buffers are safe, the value of $unload(D_n)$ is the value of $load(E_n)$ for every $n \in J_0$ for which D_n is successful.

Now to prove (11) let j be any index in J_0 such that D_j is successful. Let

$$E_m = activate(D_j).$$

Then $j \leq m$ by the Lemma 1.8. Hence $E_j \preceq E_m$. But, $E_m \prec D_j$ (by Lemma 1.5(1)), and hence $E_j \prec D_j$ follows.

Now to prove (12) assume that E_{i+N} is defined (i.e., $i + N \in I_0$) and E_{i+N} is successful. Then D_i is defined, successful and $D_i \prec E_{i+N}$ by the following argument. Lemma 1.5 shows that

$$D_n = activate(E_{i+N})$$

is a successful *DEQUEUE*. Then (by the Main Lemma) $i + N \leq n + N$, and hence $i \leq n$. So $D_i \preceq D_n$. But $D_n \prec E_{i+N}$, and hence $D_i \prec E_{i+N}$. $\square$

1.3. Liveness. In intuitive terms, liveness means that a *DEQUEUE* event must be successful if the buffer is non-empty, and an *ENQUEUE* must be successful if the buffer is not full. Given an event r, what do we mean by "the buffer is non-empty at r"? To clarify this, define

$$D(r) = \{X \mid DEQUEUE(X) \wedge X \prec r\}$$

and

$$E(r) = \{X \mid ENQUEUE(X) \wedge X \prec r\}.$$

So $D(r)$ is the set of all *DEQUEUE* events that precede r, and $E(r)$ is the set of all *ENQUEUE* events that precede r. The finiteness property implies that these sets are finite. Define

$$b(r) = |E(r)| - |D(r)|$$

(for any set S, $|S|$ denotes the number of elements of S). Then $b(r)$ represents the number of items enqueued but not dequeued by events that finished before r; so that the non-emptiness of the buffer at r can be expressed by $b(r) > 0$, and the non-fullness can be expressed by $b(r) < N$.

LEMMA 1.9. *For any r, $0 \leq b(r) \leq N$.*

Proof. Suppose for a contradiction that $b(r) > N$. Let E_i be the last *ENQUEUE* event X such that $X \prec r$ (why is there such an X? because $E(r)$ is certainly non-empty if $b(r) > N$, and by the finiteness property). So $E(r) = \{E_0, \ldots, E_i\}$ and

$$|E(r)| = i + 1.$$

Let $D_j = activate(E_i)$. Then $activate(E_i) \prec E_i \prec r$ and hence $D_j \prec r$. So $j + 1 \leq |D(r)|$. Hence

$$b(r) = |E(r)| - |D(r)| \leq |E(r)| - (j+1) = i - j.$$

But then $N < b(r) \leq i - j$ contradicts our main result (Lemma 1.7). In a similar way $0 \leq b(r)$ is proved. $\square$

LEMMA 1.10.   (1) *If r is a read at the wait part of a DEQUEUE execution X, and $0 < b(r)$, then r is successful.*
(2) *If r is a read at the wait part of an ENQUEUE execution X and $b(r) < N$ then r is successful.*

Proof. We leave (1) to the reader and prove (2). Suppose that the given *ENQUEUE* execution X is E_i and r is a read in X such that $b(r) < N$, and we shall prove that r is successful. This means that we must show that $in - out \neq N$ holds in X after executing r. Now $in(E_i) = i$, and $i \geq N$ (for $i < N$ E_i is immediate). Since $r \in E_i$, $E(r) = \{E_0, \ldots, E_{i-1}\}$ and hence $b(r) = i - |D(r)|$. Hence $b(r) < N$ implies $i - N < |D(r)|$, so that $D(r) \neq \emptyset$ as $i - N \geq 0$. Thus $\omega(r)$ is not the initial write on *Out*. Let D_j be the *DEQUEUE* operation containing $\omega(r)$. Then $Out(D_j) = j + 1$ and hence $Value(r) = j + 1$. Hence

$$in(E_i) - out(E_i) = i - (j + 1).$$

ENQUEUE (x : Item)	DEQUEUE(x : Item)
var in, out : $0..N$;	var in, out : $0..N$;
p : $0..N-1$;	p : $0..N-1$;
	(1) while $in - out = 0$
	do $Read(In, in)$;
(1) while $in - out = N$ (mod $N+1$)	(2) $unload(buf[p], x)$;
do $Read(Out, out)$;	(3) $p := p + 1$ (mod N);
(2) $load(buf[p], x)$;	(4) $out := out + 1$ (mod $N+1$);
(3) $p := p + 1$ (mod N);	(5) $Write(Out, out)$
(4) $in := in + 1$ (mod $N+1$);	
(5) $Write(In, in)$	

FIGURE 7.32. *ENQUEUE/DEQUEUE* with bounded numbers.

On the other hand $\omega(r) \in D_j$ implies

$$D(r) = \{D_0, \dots, D_j\}.$$

So $b(r) = i - (j + 1)$. But $b(r) < N$ is assumed, and hence $i - (j + 1) < N$ proving the required inequality $in(E_i) - out(E_i) \neq N$. $\square$

We can conclude now that $I_0 = J_0 = N$ and hence that all events are successful. We shall prove the following.

(13) $j \in J_0$ and D_j is successful $\Rightarrow j + N \in I_0$ and E_{j+N} is successful.

(14) $i + N \in I_0$ and E_{i+N} is successful $\Rightarrow$

each $D_{i+\ell}$ for $1 \leq \ell \leq N$ is successful.

Since $E_0, \dots, E_{N-1}$ are successful (immediate) and D_0 is successful, it follows from (13) and (14) that $I_0 = J_0 = N$. Indeed, if J_0 is finite and $j_0 = \max(J_0)$, then $j_0 \geq 1$ and D_{j_0-1} is successful. Hence by (14), for $i = j_0 - 1$, D_{i+1} is successful! This contradiction shows that $J_0 = N$, and hence (11) implies that $J_0 \subseteq I_0$ and $I_0 = N$.

2. Bounded sequence numbers

We describe in this subsection how the unbounded sequence numbers can be replaced by bounded ones. Specifically, we will use only $N+1$ coordinating numbers for in and out, and retain the N-place buffer with its (mod N) circulating variable p. We only indicate the necessary changes (not too many are needed).

The serial registers In (from *Producer* to *Consumer*) and Out (from *Consumer* to *Producer*) carry now values in $[0, \dots, N]$ where $N \geq 1$. Initially, In and Out have value 0.

buf is an N-place array of "cells" $buf[0], \dots, buf[N-1]$, which carry data values of type *Item* just as before.

The local variables, in, and out are in $0..N$, and are initially 0. The local variable p has values in $0..N-1$, and is initially 0.

Let $\mathcal{H}$ be a higher-level system execution execution resulting from an application of the protocol. The set of events H is defined as before. It consists of the read/write events on the registers, the load/unload events on the buffer cells, and the higher-level $DEQUEUE$ and $ENQUEUE$ operation executions, obtained by grouping lower level events.

The successful higher-level $ENQUEUE$ and $DEQUEUE$ events in $\mathcal{H}$ are enumerated as $\langle E_i : i \in I_0 \rangle$ (in $Producer$), and $\langle D_j : j \in J_0 \rangle$ (in $Consumer$), where I_0, J_0 can be a finite initial segment or the complete set of the natural numbers.

The values of the functions $in(X)$, $out(X)$, $In(X)$, and $Out(X)$ are defined as before on successful operation executions X. So, for example, $in(X)$ is the value of variable in after executing line 1 (since X is successful, the while loop is terminating). The value of in as it is written on register In in line 5, in case X is an $ENQUEUE$ operation execution, is denoted $In(X)$. Out is similarly defined.

The function $activate$ is defined as previously, and Lemmas 1.5 and 1.6 still hold with no change in the proof. In particular (1.6):

(1) If $E_m = activate(D_i)$, then $load(E_m) \prec unload(D_i)$.
(2) If $D_n = activate(E_k)$, then $unload(D_n) \prec load(E_k)$.

NOTATION 2.1. *For any integer k let $\lceil k \rceil_{N+1}$ be that integer $j \in [0, \dots, N]$ such that $j = k \pmod{N+1}$.*

There are some changes in Lemma 1.3, because of the use of $\pmod{N+1}$ operations in the bounded protocol. So we reformulate the lemma (and leave its simple proof as an exercise).

LEMMA 2.2. *the following holds in $\mathcal{H}$.*

(1) $in(E_0) = 0$; $out(D_0) = 0$. *And for any E_i*

$$in(E_i) = \lceil i \rceil_{N+1}, \; In(E_i) = \lceil i+1 \rceil_{N+1}.$$

For any D_j

$$out(D_j) = \lceil j \rceil_{N+1}, Out(D_j) = \lceil j+1 \rceil_{N+1}.$$

(2) *If D is a successful $DEQUEUE$, then*

$$in(D) \neq out(D).$$

If E is a successful $ENQUEUE$, then

$$in(E) - out(E) \neq N \pmod{N+1}.$$

The lemmas corresponding to Lemmas 1.7 and 1.8 can be proved as before. For example, let's prove Lemma 1.8 which was left then as an exercise:

LEMMA 2.3. *If $E_i = activate(D_j)$ then $j \leq i$.*

procedure *ENQUEUE* (x : *Item*)	procedure *DEQUEUE* (x : *Item*)
var *in*, *out* : N;	**var** *out* : N;
begin	*a* : **boolean**
(1) $P(S)$;	**begin**
(2) *Read*(In, in);	(1) **repeat** *Read*(*Arrived*[$\lceil out \rceil_N$], a)
(3) **if** *in* $\geq N$ **then repeat**	**until** $a = $ **true**;
Read(*Out, out*)	
until $in - out \neq N$;	(2) *unload*(*Buf*[$\lceil out \rceil_N, x$);
(4) *Write*($In, in + 1$);	(3) *Write*(*Arrived*[$\lceil out \rceil_N$, **false**);
(5) $V(S)$;	(4) *out* := *out* + 1;
(6) *load*(*Buf*[$\lceil in \rceil_N$]);	(5) *Write*(*Out, out*)
(7) *Write*(*Arrived*[$\lceil in \rceil_N$], **true**)	**end**
end	

FIGURE 7.33. Multiple Producers Protocol. Initial value of *Out* and *In* is 0. Initial value of *Arrived*[k] is *false*. The initial value of the semaphore S is 1.

Proof. By induction on $j \geq 0$. For $j = 0$, clearly $j \leq i$ since $i \geq 0$ whatever E_i is. Now assume that the lemma holds for all values less than j but not for j. That is $E_i = activate(D_j)$ but

$$i < j.$$

Hence the lemma holds for i by the inductive assumption. Thus, if $E_{i_0} = activate(D_i)$, then $i \leq i_0$ holds. Since $i < j$, $activate(D_i) \preceq activate(D_j)$ by monotonicity. That is, $E_{i_0} \preceq E_i$. So $i_0 \leq i$. Yet $i \leq i_0$, and hence $i_0 = i$ (and $E_{i_0} = E_i$). Apply monotonicity again to $D_i \prec D_{i+1} \preceq D_j$, and conclude that

$$E_i = activate(D_{i+1}).$$

Lemma 1.5(4) and the last lemma imply that $in(D_{i+1}) = In(E_i) = \lceil i + 1 \rceil_{N+1}$. Also $out(D_{i+1}) = \lceil i + 1 \rceil_{N+1}$. Hence $\lceil in - out \rceil_{N+1} = 0$ holds in D_{i+1}. But since $0 \leq in, out \leq N$, this equality is possible only if $in - out = 0$, which can never be in a successful *DEQUEUE*! $\square$

The proof of the safety property is as before word for word.

EXERCISE 2.4. *This exercise contains a protocol for several producers (the number of which may be not known in advance) and a consumer. The protocol uses a circular buffer containing N (some fixed number) safe buffer cells* $Buf[0], \ldots, Buf[N - 1]$. *To coordinate the producers, a semaphore S is used, and to execute an ENQUEUE operation any producer has to pass a P(S) operation. An array of boolean registers* $Arrived[0], \ldots, Arrived[N - 1]$ *is used by the processes to coordinate the accesses of the consumer. Prove the correctness of the protocol. (We shall return to this protocol in the next chapter.)*

PART 3

Message Communication

This part is about messages, channels, and networks. From this vast area two problems are chosen in order to illustrate how the models described in the first part apply.

Chapter 8 specifies channels (which connect two processes) and networks (which are used in group communication). Causal ordering is also defined there and the question of preserving causality is discussed. Since the channels may have a limited capacity, the problem of regulating the message flow is again a producer/consumer problem, and Chapter 9 describes the "sliding window" solution to this problem. The protocol described there reorders the incoming messages so that the users perceive it as a FIFO channel. In contrast to binary point-to-point communication, group communication means that the processes can broadcast messages to a group of processes in a single operation. The generalization of the FIFO ordering to the setting of a group communication is the subject of Chapter 11; we present there a variant of the ToTo protocol of Dolev, Kramer, and Malki [14] which ensures a uniform ordering of message delivery.

8
Specification of channels

Different properties of channels such as message loss, duplication, and capacity are specified. A reordering protocol illustrates some of these notions. The Sliding Window Axioms are described, and a surprising application is found.

If a register resembles a blackboard, channels and networks are rather like a mail system that distributes letters. The writing on the blackboard can be read several times until it is rewritten, but once the letter is picked up the mailbox is emptied. The writing is immediate, but the sending of letters is an elaborate process with several intermediate stages.

Registers and buffers are *immediate* data devices: This means (for example) that after executing a $load(b, x)$ instruction the loader is ensured that the value of x is loaded onto buffer b and any subsequent unload execution returns x. In contrast, when a message is sent on a channel or a network no such guarantee is given. Not only the sender does not know how long it takes to the message to arrive, but he cannot even be sure that it will arrive.

Consider the following three problems from which message systems may suffer:

(1) systems may lose messages,
(2) systems may duplicate messages (i.e., deliver twice or more the same message to the same receiver),
(3) systems may deliver messages not in the order they were sent.

There are other possible deficiencies: messages may be corrupted (transferred with errors) or their delivery may be retarded, but we shall not consider these problems here. An "orderly operating system" is one that suffers from none of these problems (we also say that this system is FIFO—first in first out). A channel can have a "capacity", which is a finite number N representing the number of messages that can be sent without blocking the channel even if no receiver is active. Our aim is to give formal definitions for these notions, and to illustrate them with a simple protocol.

1. Channels

Two sets of instructions are used here by procedures that access a channel: *Send/Receive* and *ToChannel/FromChannel*. We use the first pair typically for an orderly operating channel, and the second pair for one that may malfunction. (See figures 8.36 and 8.37 for an example where both sets of instructions are employed.)

There are two extremities to the channel: Station A and Station B. These are taken as predicates: $A(e)$ for example, says that e is an event that occurs in extremity A. We do not care about the intermediate steps a message may take to cross the channel. It is conceivable that the channel is in fact part of a complex system; messages may be sent via different routes to their destination (which may be a reason for disordering and duplication). Anyhow, a message sent at A is supposed to be destined to B and vice versa. The first-order language that we are about to define uses predicates *send/receive* to denote executions of instructions *ToChannel/FromChannel* (or instructions *Send/Receive*). So, for example, formula *receive*$(e) \wedge B(e)$ says that e is a receive event executed at extremity B.

For simplicity, we assume here that messages are only sent from A to B. The reader will have to make the obvious modifications, when in the following chapter we shall use a bi-directional channel. So we make now the assumption that all send events fall under predicate A, and all receive events under B.

When modeling a channel we must decide what exactly constitute a *Send/Receive* event. Is a send event included in the temporal interval that starts with the invocation of the *Send* operation and ends with the return? Or is the *Send* extended to include the subsequent activity of the system, which may continue long after the calling process has resumed its activity? If the first alternative is chosen, then the send events are serially ordered (assuming that *Sender* is a serial process), but if the extended interpretation is adopted, then send events may overlap. Indeed, the process may call *ToChannel* twice, and the system may be still occupied with the first call even after everything connected with the second call has terminated.

How about receive events? If a program executes a *FromChannel* instruction and this operation terminates, does the receive event start from the moment that the process invoked the *FromChannel* instruction or from the moment that the system made the incoming message available? Suppose, for example that we choose to represent a receive event by a temporal interval that corresponds to the call–return moments, and let r be that event. Let s be the corresponding send event. Then $s \prec r$ does not necessarily hold. It is indeed conceivable that at the moment that the receiver invoked *FromChannel*, namely at the start of r, the sender is doing something else and only after a delay it decides to send the message returned finally by r. Since it is conceptually simpler to think that $s \prec r$, I prefer to model *Receive* with restricted intervals.

Another important issue is whether every send/receive event is terminating. If message loss is possible, then it is reasonable to assume that every send event is terminating: in termination, the system "says" to the sender that it will take care of the message and do its best to transfer it to the other extremity of the channel.

No guarantee is made that the message ever gets there, and hence it is reasonable to assume that the sending (as a local operation) is always terminating. However, when a no-loss channel is modeled, it makes sense to allow for non terminating send events, because otherwise a channel of infinite capacity must be assumed. Indeed, the receiver may be inactive for arbitrarily long periods during which the sender may multiply its messages beyond any possible finite capacity to buffer the undelivered messages. Since we are considering mainly no-loss channels, we also allow for non terminating send events.

Now for the receive events there is another question: What should be the result of a receive operation executed when the channel is empty? There seems to be three possible answers: (1) To ensure that this never happens, i.e., that any receive operation is prompt by some signal from the channel (to the user) saying that a message is waiting there. (2) To allow for the "empty value" to be a possible return value of a receive event (in this case we probably should require that receives are always terminating, and then these operations can be used to test the status of the channel). (3) To require that a receive operation must return with some message that was sent, but to allow for non terminating receive events when the channel is (or seems to be) empty. We adopt the third answer because of its simplicity.

We say that a receiver is attentive if it repeatedly executes *FromChannel* (or *Receive*) instructions. For example, if its program is just

$$\textbf{repeat\_forever} \quad FromChannel(x).$$

Supposing that the channel losses no messages, we expect that if the receiver is attentive in some system execution, then all send events in that system execution are terminating. Since we do not want to speak about programs at this stage, we shall define attentiveness of a channel to mean that either it contains a non terminating receive event, or else infinitely many receive events. That is, a channel with only finitely many receive events which are all terminating is not attentive.

To each terminating receive event e there correspond a sent event denoted $\gamma(e)$. Thus γ is a function from terminating receive events into send events. Properties of γ reflect properties of the channel. For example, "γ is one-to-one" says that there is no duplication: no two receive events correspond to the same send. By excluding a send event s from the range of γ we can say that s was never received.

After these motivating remarks, we write down (in Figure 8.34) the signature of the language used here to specify channels.

DEFINITION 1.1 (RAW CHANNEL, ATTENTIVE CHANNEL). *Given a system execution S for this channel's signature, we say that S models a raw channel iff the properties of Figure 8.35 hold in S.*

We say that a channel is attentive (at extremity B) in S iff either there exists in S a non terminating receive event or else infinitely many receive events.

DEFINITION 1.2 (NO DUPLICATION AND ORDER PRESERVATION). *We say that a channel is with no duplication (in S) if and only if γ is one-to-one. That is, no two receive events are caused by the same send.*

Channel's signature

(1) There are two sorts: Events and Data.
(2) The following unary predicates are defined on Events: *Terminating*, A, B, *send*, *receive*. (*Terminating* is a standard predicate, and so is the binary precedence relation $\prec$. A and B refer to the channel extremities.)
(3) Two (partial) function symbols. γ : Events $\rightarrow$ Events, and *Content* : Events $\rightarrow$ Data.

FIGURE 8.34. Channel's signature.

Channel's Axioms

(1) *send* is serial: any two send events are $\prec$ - comparable:

$$\forall e_1, e_2 \ (send(e_1) \wedge send(e_2) \wedge e_1 \neq e_2 \rightarrow e_1 \prec e_2 \vee e_2 \prec e_1).$$

Also *receive* is serial. (However a send and a receive may be concurrent.)
(2) The domain of γ is the set of all terminating receive events (in station B), and its range consists of (terminating) send events (at A); the value of a terminating receive event is the value of the corresponding send event:

$$\forall e \ \ [receive(e) \wedge Terminating(e) \rightarrow send(\gamma(e)) \wedge \gamma(e) \prec e \wedge \\ Content(e) = Content(\gamma(e))].$$

(The requirement $\gamma(e) \prec e$ implies that $\gamma(e)$ is terminating.)

FIGURE 8.35. Channel's axioms.

A channel is said to be order preserving *if and only if γ is order preserving. That is, for every terminating receive events e_1, e_2, $e_1 \prec e_1$ implies $\gamma(e_1) \prec \gamma(e_2)$.*

Since receive events are serially ordered, order preservation implies no duplication.

DEFINITION 1.3 (LOST SEND EVENT). *A terminating send event e is said to be* lost *if and only if the channel is attentive but e is not in the range of γ.*

This definition may need some explanation. We think of two processes that operate independently at the two extremities of the channel; we call them Sender and Receiver. Suppose that for some reason Receiver does not execute at all, or stops after a few steps. If in such a case there are more send then receive events, then some send events will never be received even though we cannot really blame the channel for that. So, not being in the range of γ is a necessary but not a sufficient reason for saying that the channel has lost the message, and we need the additional information that the channel is attentive in order to deduce that a sending event not in the range of γ is lost.

This leads to the following

DEFINITION 1.4 (NO LOSS). *Let S be a system execution for the channel signature defined above. We say that the channel is* non-lossy *in S if the following implication holds:*

If the channel is attentive, then γ is onto the set of send events.

Observe that in a non-lossy channel there cannot be both a non terminating send event and a non terminating receive event. In fact, if there is a non terminating receive event then the channel is attentive and hence every send event is in the range of γ and is thus terminating. Attentiveness will be further clarified in the next section.

DEFINITION 1.5 (ORDERLY OPERATING). *An attentive channel is said to be* orderly operating *if it is non lossy and order preserving. We also say that such a channel is* FIFO. *That is, an attentive channel is* FIFO *if γ is order preserving and onto the send events.*

1.1. Capacity of a channel. The capacity of a channel is the number of messages that the channel can normally hold, before its degradation. We think of a channel as some kind of a buffer which the sender fills and the receiver discharge. The capacity is the maximal number of messages that the channel can hold without being discharged. Formally, we give the following definition.

Let S be a system, consisting of system executions for the channel signature. Let us agree that in the following definition variables s and $s_1, \ldots, s_N$ only vary over send events.

DEFINITION 1.6. *We say that a non-lossy channel has "capacity N in S" if and only if the following two properties hold in any S in S:*

(15) $\forall s[\neg Terminating(s) \longrightarrow \exists s_1, \dots, s_N \bigwedge_{i \neq j} (s_i \neq s_j) \wedge$

$$\bigwedge_{1 \leq i \leq N} (s_i \prec s \wedge s_i \text{ is not in the range of } \gamma)]$$

(16) $\forall s_0, \dots, s_N [\bigwedge_{0 \leq i < N} s_i \prec s_{i+1} \wedge Terminating(s_N) \longrightarrow$

$$\exists r (receive(r) \wedge r \prec Right\_End(s_N) \wedge \gamma(r) = s_0)].$$

In words, the first formula says that a send event is non terminating only if the channel is full, namely there are N preceding send events that are not received by B. This formula alone will not suffice: it does not exclude for example the possibility that all send events are terminating and there is no receive event at all. So, a stricter motion of capacity is required. One possibility is to transform the implication into a bi-conditional (if and only if) connective, but here we have an even stronger requirement.

Formula (16) says that in a non-lossy channel of capacity N the sender knows for sure that its kth message was received when the $(k + N)$th send operation has terminated. For example, after the Nth send operation has ended it is still possible that all messages are buffered by the channel and none has been received by the receiver. However, since $N + 1$ messages are never buffered, after the $(N + 1)$th send has ended it must be the case that the first send has been received. In this sense, smaller capacity provides greater assurance. For example, in a channel of capacity 1, the sender knows after the second send operation that the first receive operation has terminated.

2. A redressing protocol

I think that the notion of attentive channel needs a clarifying example, and the Redressing Protocol is brought for this purpose. Since the Sliding Window protocol of the following chapter is more involved and is described there in all detail, we shall be very sketchy here and give no proofs.

Receiving messages in the order they were sent is a desirable property, and the users of a channel may agree to a certain slowdown of the system in exchange for an assurance that the channel is FIFO. A very simple protocol can restore the original ordering of the messages: The sender (located at extremity A of the channel) attaches natural numbers (also called timestamps or sequence numbers) to its messages in increasing order, and the receiver (at extremity B of the channel) buffers the incoming messages and delivers a message only if all messages with smaller sequence numbers have already been delivered. An infinite buffer is assumed here, which considerably simplifies the protocol. So we assume safe buffers $Buf[i]$ for $i \in \mathbb{N}$. In the following chapter this infinite array is replaced by a bounded buffer.

We assume that the given channel (which we also call the raw channel) is a no-loss channel in which disordering is possible.

procedure *Producer* begin repeat_forever (1) Produce_item(a); (2) *Send*(a); end	procedure *Consumer* begin repeat_forever (1) *Receive*(b); (2) Consume_item(b); end
procedure *Handler* begin repeat_forever *DataArr* end	
protocol Redressing concurrently do *Producer, Consumer, Handler* od	

FIGURE 8.36. *Producer/Consumer* sending and receiving items.

The general setting of our protocols is described in Figure 8.36. This is a producer/consumer protocol with two changes compared with the generic *Producer/Consumer* protocol of Chapter 6:

(1) Instructions *Send/Receive* are used here instead of *ENQUEUE/DEQUEUE*.
(2) *DataArr* is repeatedly executed by *Handler* which is a "background" procedure that collects the incoming messages and stores them for the consumer. The users of the channel are, in a sense, unaware of this service rendering procedure.

Operations Produce_item, and Consume_item are assumed to always terminate (for simplicity). *Item*, the type of variables a and b, is the data type produced or consumed.

Procedures *Send/Receive/DataArr* are described in Figure 8.37).

Let S be a system execution of this redressing protocol. We want to prove that every *FromChannel* event and every *ToChannel* event in S are terminating and that there are infinitely many events in each type. The channel is assumed to be non-lossy. By Definition 1.4 this means that if the channel it attentive, then γ is onto the set of *ToChannel* events.

First we prove that the channel is *attentive*. Considering the definition, we must prove that if every *FromChannel* is terminating, then there are infinitely many such events. But looking at procedure *DataArr*, we see that line (1) (containing instruction "*FromChannel*(r)") is the only possibly nonterminating instruction in this procedure, since loading and writing instructions are always terminating. As *Handler* is repeatedly executing *DataArr*, an infinite sequence of *FromChannel* events results (from our assumption that every *FromChannel* is terminating). Thus the channel is attentive. Now, since we assume a no-loss channel, γ is onto the set of *ToChannel* events, and as $t = \gamma(r)$ implies $t \prec r$, every *ToChannel* event is terminating. In turn, this implies that there are infinitely many *ToChannel* events, and correspondingly infinitely many terminating *FromChannel* events (i.e. all the γ pre-images).

To enhance the contrast of this argument suppose that *DataArr* were the

procedure *Send*(*a* : *Item*)	procedure *DataArr*
var *s* : *DataFrame*;	var *r* : *DataFrame*
next_to_send: *SequenceNr*	begin
begin	
	(1) *FromChannel*(*r*);
(1) *s*.info := *a*	(2) *load*(*Buf*[r.tag], r.info);
s.tag := *next_to_send*;	(3) *Write*(*Arrived*[r.tag], true)
(2) *next_to_send* := *next_to_send* + 1;	end
(3) *ToChannel*(*s*)	
end	

procedure *Receive*(*b* : *Item*)
begin
(1) *wait-until Arrived*[*frame_expected*];
(2) *unload*(*Buf*[*frame_expected*], *b*);
(3) *frame_expected* := *frame_expected* + 1
end

FIGURE 8.37. *DataArr* is repeated forever at *B*.   *Receive* is called by the user at *B*.   *frame_expected* is initially 0. *next_to_send* is initially 0.

following procedure

```
while i < 5 do
(1) FromChannel(r);
(2) i = i + 1
```

Then (assuming i is initialized to 0) only five *FromChannel* events can be executed, and the channel is not attentive. In this setting, one cannot deduce that γ is onto the *ToChannel* events, and conceivably some messages were never received.

The correctness proof of this Redressing Protocol will not be further pursued since my main aim is to replace the unbounded buffer array with a finite array of buffer cells, and to prove the correctness of the resulting more complex protocol. This will be done in the following chapter, and my aim in bringing the Redressing Protocol here is to illustrate *attentiveness*.

3. Sliding Window Axioms

The following chapter is devoted to a detailed description of a sliding window protocol, but this section describes abstract properties of such protocols. (The reader who prefers to see the concrete before the abstract can proceed directly to the following chapter.) First we describe the language for the axioms.

Instead of *send/receive* and γ used in the first section, we use in this new context *SEND/RECEIVE* and Γ, and an additional predicate *ACK* to describe acknowledgment events. The language contains an enumeration $\langle S_i \mid i \in I_0 \rangle$ of the *SEND* events, and $\langle R_j \mid j \in J_0 \rangle$ of the *RECEIVE* events. Formally, we should write $S(i)$ and $R(j)$ (viewing S and R as functions) but the subscript

Sliding Window Axioms

---

(1) $\langle R_j \mid j \in J_0 \rangle$ is an enumeration in increasing order of all *RECEIVE* events, where J_0 is an initial segment of N. For $i < j$ in J_0, $R_i \prec R_j$. Either J_0 is N, or else it is finite, non-empty and then R_{j_0} is nonterminating, where $j_0 = \max(J_0)$.

(2) $\langle S_i \mid i \in I_0 \rangle$ is an enumeration of all *SEND* events. It is not necessarily serial, but $S_i \prec S_j$ implies $i < j$. Either $I_0 = $ N and then each S_i is terminating, or else I_0 is a finite, non-empty initial segment and then (only) S_{i_0} is nonterminating, where $i_0 = \max(I_0)$.

(3) *activate* : *SEND* $\to$ *RECEIVE* satisfies the following.

   (a) *activate* is defined on the terminating *SEND* events S_i with $i \geq N$.

   (b) If $R = activate(S)$ then $R \prec S$.

   (c) If $R_j = activate(S_i)$ then $i \leq j + N$.

(4) Γ : *RECEIVE* $\to$ *SEND* is defined on the terminating *RECEIVE* events with values that are *SEND* events.

   (a) $\Gamma(R) \prec R$, for every terminating *RECEIVE* event R.

   (b) Γ is one-to-one.

   (c) If $\Gamma(R_j) = S_i$ then $j = i \pmod N$.

(5) *ACK* is a linearly ordered set of events. *ack* : *ACK* $\to$ N satisfies:

   (a) If R_j is terminating then there is an *ACK* event A such that $R_j \preceq A$, and for any such *ACK* event $j + 1 \leq ack(A)$.

   (b) If S_i is nonterminating, then $i \geq N$ and there is an *ACK* event A_0 such that, for every *ACK* event A, if $A_0 \preceq A$ then $ack(A) \leq i - N$.

(6) Suppose that R_{j_0} is non-terminating (by (1) this implies that $j_0 = \max(I_0)$). If $j_0 \in I_0$ and S_{j_0} is terminating then $j_0 \geq N$ and

$$\neg(R_{j_0 - N} \prec S_{j_0}).$$

FIGURE 8.38. The Sliding Window Axioms

notation is more common for this purpose. The domains I_0 and J_0 are initial segments of N (finite or infinite). Unlike the raw channel axioms, we do not assume here that the *SEND* events are serially ordered. Intuitively this results from our understanding of a *SEND* event as a complex event that includes not only the local sending but the arrival of the message to the other end of the channel as well.

Three functions are used in the specification: *activate*, Γ, and *ack*. N is a constant (interpreted as an integer ≥ 1).

The Sliding Window Axioms are in Figure 8.38. Some words to motivate these axioms. The channel has a bounded capacity $N \geq 1$, and the acknowledgment messages (from the receiver to the sender) are used to regulate the sending. The sender can send the first N messages $S_0, \ldots, S_{N-1}$ without hesitation, since the channel can absorb the first N messages. For any subsequent messages, the sender must obtain a confirmation from the receiver. These confirmations are the *ACK* events. Acknowledgment messages tell the sender the number of messages so far received. If A is an acknowledgment event (which we write as

$ACK(A))$ then $ack(A) \in \mathbb{N}$ is the acknowledgment number of A. If $R_j \preceq A$ then at least $j + 1$ messages have been received with event A (namely $A_0, \ldots, A_j$) and thus $ack(A) \geq j + 1$ (which is 5(a)). Notice that I write $R_j \preceq A$ rather than $R_j \prec A$, because at least in one example there is a possibility that receive events are identified with acknowledgment events.

The function $activate$ relates the sending to the confirming $RECEIVE$ events. If S_i, for $i \geq N$, is terminating, then $activate(S_i)$ is the $RECEIVE$ event on which S_i bases its assumption that the channel is not full. If $R_j = activate(S_i)$ then S_i "knows" for sure that R_j has terminated. This information is reaching S_i via an ACK event A with $ack(A) = j + 1$. If $R_j = activate(S_i)$ then $i \leq j + N$ (that is property 3(c)). It follows, in fact, that i is in the window $j + 1 \leq i \leq j + N$.

Γ is the return function. Namely, $\Gamma(R_j) = S_i$ means intuitively that the message sent by S_i was received by R_j. We shall prove that $\Gamma(R_i) = S_i$ (and $I_0 = J_0$), but axiom 4(c) only gives $j = i \pmod{N}$.

If a sending event S_i fails to terminate, despite the fact that it has observed infinitely many acknowledgment messages, it must be the case that these messages gave evidence for only $i - N$ receive messages, and thus the ith message would fill the channel beyond its capacity.

It seems harder to justify in general, convincing terms the sixth axiom. The assumption made in axiom 6 is that J_0 is finite and R_{j_0} is nonterminating. Since sending and receiving of the first N messages are necessarily terminating (by the channel having capacity N), clearly $j_0 \geq N$. Axiom 6 excludes the possibility that $R_{j_0 - N} \prec S_{j_0}$. The reader may recall that Axiom 6 is, in a slight disguise, a formulation of Theorem 3.3 from Chapter 5 concerning the KanGaroo/LoGaroo Protocol.

In the following lemmas we derive some consequences of these axioms.

LEMMA 3.1. *For every $k \in J_0$ if R_k is terminating then $\Gamma(R_k) = S_k$.*

Proof. The proof is by induction on $k \in \mathbb{N}$. Let

$$(17) \qquad\qquad S_i = \Gamma(R_k).$$

Then S_i is terminating by 4(a), and $i = k \pmod{N}$ by 4(c). If $i < k$ then induction is applied to i and hence $\Gamma(R_i) = S_i$, which contradicts the assumption that Γ is one-to-one (axiom 4(b)). Hence $i \geq k$, and we must derive a contradiction from $i > k$ in order to conclude that $i = k$. But $i > k$ implies

$$(18) \qquad\qquad i \geq k + N$$

since $i = k \pmod{N}$. So, as $i \geq N$, $R_j = activate(S_i)$ is defined, and

$$(19) \qquad\qquad i \leq j + N$$

by 3(c). Now $S_i = \Gamma(R_k)$ and $R_j = activate(S_i)$ imply $S_i \prec R_k$ and $R_j \prec S_i$ (by 4(a) and 3(b)). Hence $R_j \prec R_k$, and $j < k$ (because the enumeration is in increasing order). So $i \leq j + N < k + N$, in contradiction to (18). $\square$

LEMMA 3.2. $J_0 = I_0$.

Proof. Suppose first that $k \in J_0$ and we shall prove that $k \in I_0$. If R_k is terminating, then $S_k = \Gamma(R_k)$ was established above and then in particular $k \in I_0$. If on the other hand R_k is nonterminating, then $k = \max(J_0)$. If $k = 0$ then $k \in I_0$, since I_0 is a non-empty initial segment. Assume $k > 0$. So R_{k-1} is terminating, and hence $S_{k-1} = \Gamma(R_{k-1})$ is terminating. Thus $k - 1 \in I_0$ and it cannot be the last index there (as S_{k-1} is terminating). Thus $k \in I_0$.

Now we prove by induction on i that $i \in I_0 \implies i \in J_0$. Assume that i is the first counterexample. Then $i \in I_0 \setminus J_0$. So J_0 is finite and if $j_0 = \max(J_0)$ then R_{j_0} is nonterminating. Since $j_0 < i \in I_0$, S_{j_0} is terminating (by the last clause of 2). By Axiom (6), $j_0 \geq N$ and

$$\neg(R_{j_0 - N} \prec S_{j_0}).$$

Yet $R_j = activate(S_{j_0})$ is defined (as $j_0 \geq N$) and $R_j \prec S_{j_0}$ (by 3(b)), and $j_0 \leq j + N$ (by 3(c)). So $J_0 - N \leq j$ and hence $R_{j_0 - N} \preceq R_j$. Hence $R_{j_0 - N} \prec S_{j_0}$, which is a contradiction. $\square$

LEMMA 3.3. $I_0 = J_0$ is infinite.

Proof. Suppose that J_0 is finite and then that R_{j_0} is nonterminating, where $j_0 = \max(J_0)$. Then $j_0 = \max(I_0)$ as well, and S_{j_0} is nonterminating. By 5(b), $j_0 \geq N$ and there is some ACK event A_0 such that for any ACK event A with $A_0 \preceq A$

$$ack(A) \leq j_0 - N.$$

Yet $R_{j_0 - 1}$ is terminating, and hence by 5(a) there is an ACK event A with $R_{j_0 - 1} \preceq A$. Since the ACK events are linearly ordered, we may take the greater of A_0 and A, and thus we may assume that $A_0 \preceq A$. Then (by 5(a)) $j_0 - 1 + 1 \leq ack(A)$ which is a contradiction to $ack(A) \leq j_0 - N$ since $N \geq 1$. $\square$

LEMMA 3.4. For every $0 \leq k < N$, the events $\{S_i \mid i = k \pmod{N}\}$ and $\{R_j \mid j = k \pmod{N}\}$ satisfy the conflict freedom property. Namely

$$S_k \prec R_k \prec S_{k+N} \prec R_{k+N} \ldots$$

That is, for all $i \in \mathbb{N}$

$$S_i \prec R_i \prec S_{i+N} \prec R_{i+N}.$$

Proof. Since $S_i = \Gamma(R_i)$, $S_i \prec R_i$ (and similarly $S_{i+N} \prec R_{i+N}$) follow from 4(a). On the other hand, if $R_j = activate(S_{i+N})$ then $R_j \prec S_{i+N}$ and $i + N \leq j + N$. Hence $i \leq j$. So $R_i \preceq R_j$ and hence $R_i \prec S_{i+N}$. $\square$

LEMMA 3.5. If $R_j = activate(S_i)$, then $j + 1 \leq i \leq j + N$.

Proof. 3(c) gives $i \leq j + N$ immediately. R_j, as any $RECEIVE$ event, is terminating (since $J_0 = \mathbb{N}$). Hence $\Gamma(R_j) = S_j$ is defined and $S_j \prec R_j$. Yet our assumption $R_j = activate(S_i)$ implies $R_j \prec S_i$. We conclude $S_j \prec S_i$, and thence $j < i$ (by (2)). $\square$

procedure $Producer_i$ **repeat_forever** (1) produce_an_item(x); (2) ENQUEUE(x);	procedure *Consumer* **repeat_forever** (1) DEQUEUE(x); (2) consume_item(x);
protocol Multiple *Producer* **concurrently do** $Producer_1, \ldots, Producer_{p_0}$, *Consumer* **od**	

FIGURE 8.39. The generic Multiple *Producer* (single) *Consumer* protocol

4. A Multiple Producer Protocol

Surely the Sliding Window Axioms appear somewhat strange at first sight, so that the decision to write them here rather than in the following chapter requires some justification. As a matter of fact, these axioms were extracted from an earlier correctness proof for the Sliding Window Protocol. That proof was rather long, and I thought to divide it into two parts by finding some higher-level "axioms" which would serve as a resting point and give some sense of direction to the first part of the proof. We have already seen correctness proofs divided into lower and higher level parts, but here we have an additional benefit of the method: that the higher-level axioms thus obtained serve for another protocol, thus not only facilitating its correctness proof, but also showing some affinities between the two protocols. Specifically, we shall use the Sliding Window Axioms here to prove the correctness of the Multiple Producer Protocol which was given as an exercise in Chapter 7. The protocol is reproduced here for the reader's convenience. The setting for the Multiple Producer Protocol is Figure 8.39 and the protocol itself is in Figure 8.40.

There are p_0 producing processes, $Producer_1, \ldots, Producer_{p_0}$ which execute the same program (but each with its own local variables of course) and a single consumer. The role of the protocol (Figure 8.40) is not only to regulate the load/unload movement, but also to prevent two producers from approaching the same buffer cell. This is done with the aid of a semaphore S: $Producer_i$ has first to obtain permission in executing $P(S)$, and only then it reads in register *In* its index number in. The role of this index is to indicate the value $\lceil in \rceil_N$ of the buffer cell *Producer* should approach. The loop condition $in - out \neq N$ must be satisfied before the producer may proceed. Before releasing the semaphore, $Producer_i$ increases by 1 the value of *In*. The other parts of the protocol resembles very much the corresponding two-process protocols described in Chapter 7.

Let S be a system execution of the Multiple Producer Protocol (figures 8.39, 8.40). We make the following assumptions:

(1) The buffer cells $Buf[i]$ for $0 \leq i < N$ are safe in S.
(2) Registers *In*, *Out* are serial in S. (In fact, the protocol itself ensures mutual exclusion of the accesses to register *In*.) The values carried by these registers are natural numbers.
(3) Registers $Arrived[i]$, $0 \leq i < N$, are serial. They are boolean.

Higher-level ENQUEUE and DEQUEUE operation executions are defined in S.

procedure *ENQUEUE* (x : *Item*)	procedure *DEQUEUE* (x : *Item*)
var *in*, *out* : N;	var *out* : N; (initially 0)
begin	a : boolean
(1) P(S)	begin
(2) *Read*(*In*, *in*);	(1) **repeat** *Read*(*Arrived*[$\lceil out \rceil_N$], a)
(3) if *in* $\geq N$ then	**until** a = **true**;
repeat	(2) *unload*(*Buf*[$\lceil out \rceil_N$, x);
Read(*Out*, *out*)	(3) *Write*(*Arrived*[$\lceil out \rceil_N$, **false**);
until *in* − *out* $\neq N$;	(4) *out* := *out* + 1;
(4) *Write*(*In*, *in* + 1);	(5) *Write*(*Out*, *out*)
(5) *V*(*S*);	end
(6) *load*(*Buf*[$\lceil in \rceil_N$]);	
(7) *Write*(*Arrived*[$\lceil in \rceil_N$], **true**)	
end	

FIGURE 8.40. Declarations of *ENQUEUE* and *DEQUEUE* for the Multiple Producer Protocol. Initial value of *Out* and *In* is 0. Initial value of *Arrived*[k] is *false*. The initial value of the semaphore S is 1.

DEFINITION 4.1. *An* ENQUEUE *execution* E *is called* alive *iff its* P_S *event is terminating. It is called* successful *if and only if it is terminating. That is:*

(1) *It is alive.*
(2) *The* if *instruction of line (3) is terminating. That is, if* $i \geq N$ *holds then the* repeat_until *loop of line (3) is terminating.*

If E is an alive *ENQUEUE* event, then E contains a read of register *In*, and the value obtained is denoted $in(E)$. If E is successful, then the value of variable *out*, after the terminating execution of line (3) is denoted $out(E)$. In case $in \geq N$ holds, then $out(E)$ is the value of the last read of register *Out* in the terminating loop. $In(E) = in(E) + 1$ is the value written in line (4).

If D is any *DEQUEUE* execution, then $out(D)$ is the value of variable *out* when D begins. (It is 0 initially.) If $j = \lceil out(D) \rceil_N$, then D contains reads of register *Arrived*[j] in executing the **repeat_until** loop of line (1). If this loop is terminating, then D is said to be *successful*. In this case

$$Out(D) = out(D) + 1$$

is the value written on register *Out* in line (5). (Here too, *successful* is equivalent to terminating, since *unload* and *Write* events are always terminating.)

Unlike the single producer protocols of Chapter 6 and 7, the *ENQUEUE* executions in our multiple producer execution S are not necessarily linearly ordered. It is possible for example that the first two *ENQUEUE* executions load their buffer cells at the same time. Of course, the intervals of P_S and V_S events are disjoint, and thus the reads of register *In* and its updatings are disjoint, but the subsequent events may well overlap. In fact, it is precisely because of this possibility that the protocol is claimed to be more efficient. So, since the *ENQUEUE* executions are not necessarily serially ordered, we cannot enumerate them as in the single producer case. This makes it slightly harder to formulate the correctness condition. In Chapter 6, the safely requirement was simply that the value of the nth *DEQUEUE* equals the value of the nth *ENQUEUE*, but here the term "nth *ENQUEUE*" is not clearly defined. So the enumeration of the *ENQUEUE* events is now part of what we want to prove. The key to this enumeration is the fact that the critical sections formed between the $P(S)$ and $V(S)$ executions are mutually excluded.

We shall make use of the result in Chapter 6 Section 5 about the Multiple Process Mutual Exclusion Protocol. Consider the producer's procedure *ENQUEUE*, and see that it has the form of a round: $P(S)$; CS; $V(S)$; B. The CS events (executions of instructions 2, 3, 4) may well be non terminating if the *ENQUEUE* executions is not successful. The B part (lines 6, 7) is always terminating. The results of Chapter 6 (Corollary 5.2) give the following.

(1) The CS events are mutually exclusive. A CS event may well be non terminating. In this case it consists of a non terminating execution of line 3.

(2) Let $\langle C_i | i \in I_0 \rangle$ be the enumeration in increasing $\prec$ order of the CS events. If every CS event is terminating, then $I_0 = \mathbf{N}$.

Let $in(C_i)$ denote the value of variable *in* as obtained in executing line 2 in C_i. Then $in(C_i) = i$ can be proved by induction on i.

Returning to our *ENQUEUE* events let E_i be that *ENQUEUE* execution containing C_i. Then $in(E_i) = i$ and the enumeration of the CS events induces an enumeration $\langle E_i \mid i \in I_0 \rangle$ of all alive *ENQUEUE* executions. Either $I_0 = \mathbf{N}$ and all E_i are successful and terminating, or I_0 is a finite initial segment of $\mathbf{N}$, and for $i_0 = \max(I_0)$ E_i is alive but non terminating. For all $i < i_0$, E_i is terminating (successful).

The following lemma is the consequence of our discussion.

LEMMA 4.2. *The values* $\{in(E) \mid E$ *is an alive ENQUEUE execution*$\}$ *form an initial segment I of* N. *If* $E \prec E'$ *are two alive ENQUEUE executions, then* $in(E) < in(E')$. *If* $E \neq E'$ *are two alive ENQUEUE executions then* $in(E) \neq in(E')$.

For every execution of the *ENQUEUE* (*DEQUEUE*) procedure we are going to define an *ENQUEUE* (*DEQUEUE*) event which is a subcollection of the corresponding execution. The full rational for this will be clarified later, but as in previous proofs the reason for not including all lower level events in the *ENQUEUE* (*DEQUEUE*) events is to highlight those lower level events that play a prominent role in the correctness proof.

(1) If X is a successful *ENQUEUE* execution for which condition $i \geq N$ holds, then X contains a successful read of register *Out*. The corresponding *ENQUEUE* event consists of this successful read, r, the load event of line 6, and the write event of line 7. Thus the successful read is the first event of the successful *ENQUEUE* event when $i \geq N$ holds.

(2) If X is alive but not successful *ENQUEUE* execution, then it contains an infinite sequence of reads of register *Out*, all satisfying condition $in - out = N$, which never terminates the loop of line 3. The corresponding ENQ event is this infinite set of read events.

(3) If X is an *ENQUEUE* execution that does not fall under (1) or (2), then the event corresponding to X is X itself.

Higher level *DEQUEUE* events are also defined. If X is a *DEQUEUE* execution, it is said to be successful in case the repeat loop of line 1 is terminating. A *DEQUEUE* execution is successful iff it is terminating, since all other lower-level events are necessarily terminating. If X is a successful *DEQUEUE* execution, then it contains a read of $Arrived[\lceil out \rceil_N]$ returning **true**, which is called the "successful read" of X. The corresponding *DEQUEUE* event is taken to be the set of lower level events formed by the successful read, the following unload event, and the subsequent two write events on $Arrived[\lceil out \rceil_N]$ and on Out (corresponding to lines 3 and 5).

If X is a non successful *DEQUEUE* event, then it contains an infinite sequence of reads of register $Arrived[\lceil out \rceil_N]$ which all return *false*, and the corresponding higher level event is this infinite set.

Now I will copy Lemma 1.3 from Chapter 7 and ask the reader to prove it in our new context.

LEMMA 4.3. *The following holds in S.*

(1) $in(E_0) = 0$; $out(D_0) = 0$. *For any E_i*

$$in(E_i) = i, \text{ and } In(E_i) = i + 1.$$

For any D_j,
$$out(D_j) = j \text{ and } Out(D_j) = j + 1.$$

(2) *If E is a successful ENQUEUE, then*

$$in(E) - out(E) \neq N.$$

The definition of $activate(X)$ is simpler here. If $E = E_i$ is a successful *ENQUEUE* event with $i \geq N$, then E contains a successful read of register Out, namely a read that obtains condition

$$i - out \neq N,$$

and then $\omega(r)$ determines $activate(E)$: If $\omega(r)$ is the initial write on Out (of value 0) then $activate(E) = \omega(r)$, but otherwise $\omega(r)$ belongs to some *DEQUEUE* event D, and we define
$$activate(E) = D.$$

If $D = D_j$ is a successful *DEQUEUE* event, then $out(D) = j$, and D contains a successful read r of register $Arrived[\lceil j \rceil_N]$. Then $\omega(r)$ is a write of value **true** on that register, necessarily in some terminating *ENQUEUE* event E and we define

$$E = activate(D).$$

We write $\Gamma(D) = E$ instead of $activate(D) = E$, and notice the resemblance of our protocol to the *KanGaroo/LoGaroo* Protocol of Chapter 5. That is, *ENQUEUE* is *KanGaroo*, and *DEQUEUE* is *LoGaroo*. Taking advantage of this resemblance, we may deduce that Γ is one-to-one (Theorem 3.2 there).

Let me copy here Lemma 1.5 from Chapter 7, and ask the reader again to prove it in this new context.

LEMMA 4.4.   (1) *For every X in the domain of activate:*

$$activate(X) \prec X.$$

(2) *If X is a successful ENQUEUE in the domain of activate(namely of the form E_i for $i \geq N$), then $activate(X)$ is a successful DEQUEUE. Similarly, if X is a successful DEQUEUE, then $activate(X)$ is a successful ENQUEUE.*

(3) *activate is monotonic: If $N \leq i \leq i'$ then*

$$activate(E_i) \preceq activate(E_{i'}),$$

and if $0 \leq i \leq i'$ then

$$activate(D_i) \preceq activate(D_{i'}).$$

(4) *Let D and E be a DEQUEUE and ENQUEUE successful events:*

$$If\ D = activate(E),\ then\ out(E) = Out(D).$$

Lemma 1.6 of Chapter 7 is valid here too and I ask the reader again to check it.

LEMMA 4.5.   (1) *If $E_m = activate(D_i)$, then $load(E_m) \prec unload(D_i)$.*
(2) *If $D_n = activate(E_k)$, then $unload(D_n) \prec load(E_k)$.*

Since the Main Lemma (namely 1.7 in Chapter 7) relied solely on the part of Lemmas 1.3 and 1.5 that are valid here too, it is valid in our context as well and we can use it here.

I would like to argue now that under a suitable translation the Sliding Window Axioms hold for the Multiple Producer Protocol. The translation is as follows. The *RECEIVE* events are the *DEQUEUE* events, and the *SEND* events are the alive *ENQUEUE* events. Thus we have $R_j = D_j$ for $j \in J_0$, and $S_i = E_i$ for $i \in I_0$.

The function *activate* is defined on the terminating *SEND* (i.e., *ENQUEUE*) events S_i for $i \geq N$. The function $\Gamma : DEQUEUE \rightarrow ENQUEUE$ is just the function *activate*. The *ACK* events are the initial write on *Out* and the terminating *RECEIVE* (*DEQUEUE*) events. If A is such a terminating *DEQUEUE* event, then $ack(A) = Out(A)$ is the value written on register *Out*. Now we prove the Sliding Window Axioms (leaving some detail to the reader).

(3)(c): If $R_j = activate(S_i)$ then $i \leq j + N$. This is exactly the Main Lemma.

(4)(b): Γ is one-to-one. Γ is in fact *activate* as it is defined on the successful *DEQUEUE* events. That Γ is one-to-one is an application of the *KanGaroo/LoGaroo* Protocol and specifically of Theorem 3.2 in Chapter 5.

(4)(c): If $\Gamma(R_j) = S_i$, then R_j contains a successful read r of $Arrived[\lceil j \rceil_N]$, and hence $\lceil j \rceil_N = \lceil i \rceil_N$.

(5)(a): Suppose that R_j is a terminating *DEQUEUE* event. Then R_j is itself an *ACK* event, and $ack(R_j) = Out(R_j)$ by definition. Hence $ack(R_j) = j + 1$. Now if A is any *ACK* event such that $R_j \preceq A$, then $Out(R_j) \leq Out(A)$, and hence $j + 1 \leq ack(A)$.

(5)(b): Suppose that S_i is a non terminating *SEND* event, that is an alive but non terminating *ENQUEUE* event. Then $i \geq N$ and S_i consists of an infinite set of read events of register *Out*, all establishing condition $in - out = N$. Since the value of in is $in(E_i) = i$, we have that each of these reads returns the value $out = i - N$. This shows

(a) that there is an *ACK* event (perhaps the initial write on *Out*) such that $ack(A_0) = i - N$, and

(b) for every *ACK* event A there is an *ACK* event A_1 such that $A \preceq A_1$ and $ack(A_1) = i - N$.

Hence $ack(A) \leq i - N$.

(6): This is essentially Theorem 3.1 of Chapter 5.

Now we are allowed to use the consequences of the Sliding Window Axioms. In particular $I_0 = J_0 = N$ and the events $\{E_i \mid i = k \pmod{N}\}$ and $\{D_j \mid j = k \pmod{N}\}$ satisfy the conflict freedom property. As $Buf[k]$ are safe, this implies that the value of $unload(D_j)$ equals the value of $load(E_j)$ for every $j \in N$.

9

A sliding window protocol

We assume here a channel that connects the Consumer to the Producer. The channel is a raw communication device that may disorder messages, and the problem is therefore to devise a protocol that implements an orderly operating channel. Unlike the Redressing protocol of the previous chapter, the buffers here are bounded and acknowledgment messages regulate the message flow. The protocol is proved to satisfy the Sliding-Window axioms. The chapter contains so many local definitions that a special index (page 188) is created.

We describe in this chapter another *Producer/Consumer* protocol that is appropriate when a raw channel connects the two processes.

The basic idea of the *Sender/Receiver* protocol is simple. The *Sender* attaches tag numbers to the messages sent by *Producer*, so that the n-th message is tagged with tag number $\lceil n \rceil_N$. These tag numbers are used by the receiving process to correctly order the incoming messages even if they arrive out of order. The receiving process maintains an array of N buffer cells $Buf[0], \ldots, Buf[N-1]$, and each incoming frame with tag k is buffered in $Buf[k]$ until it is delivered in correct order to the Consumer. Similar protocols can be found in almost any networking textbook (e.g. Tanenbaum's "Computer Networks", Chapter 4), but as our main aim is to study *concurrency*, we write our protocols in a somewhat different way, so that any two instructions that can be executed concurrently are written in separate, concurrent procedures. Whenever two instructions are required to be executed in sequence, then this assumption is used in the correctness proof.

The Sliding Window Protocol (Figure 9.41) concurrently combines five procedures:

> *Producer, Consumer, Handler, Pitcher,* and *Catcher.*

These procedures invoke five operation procedures:

> *Send, Receive, DataArr, PitchAck,* and *CatchAck,*

164

which are described (declared) in Figure 9.45.

The channel has two extremities called Stations A and B. *Producer* is located at Station A and *Consumer* at Station B. The five procedures comprising the Sliding Window Protocol and the operation procedures that they invoke are logically divided into two sets of processes: *Sender* and *Receiver*, which are distributed in Stations A and B as follows.

The *Sender*'s procedures:	The *Receiver*'s procedures:
(1) At Station A: *Producer*, which invokes *Send(a)*. (2) At Station B: *Handler*, which invokes *DataArr*.	(1) At Station B: *Consumer*, which invokes *Receive(b)*. (2) At Station B: *Pitcher* which invokes *PitchAck*. (3) At Station A: *Catcher* which invokes *CatchAck*.

Sender contains *Send* and *DataArr*. *Receiver* contains procedures *Receive*, *PitchAck*, and *CatchAck*. Notice how *Sender* and *Receiver* are both distributed across the two ends of the channel. *Sender*, for example, contains *Send* at Station A and *DataArr* at B. It may seem strange at first sight to consider a procedure at Station B to be part of the sender, but I view *DataArr* as some kind of agent of *Sender* which is located at the other end of the channel and whose aim is to ensure that the frames are deposited at the right buffer cell.

Similarly, though *CatchAck* is physically at Station A, it is part of *Receiver*. It obtains the acknowledgment messages from the channel and then updates register *AckReceived*.

Conceptually, we have here three layers:

(1) The users' layer contains procedures *Producer* and *Consumer*. The producer cycles through producing and sending operations, and the consumer through receiving and consuming. We assume that Produce_item and Consume_item are terminating external operations; a and b are variables of type *Item* which is the information unit produced and consumed.

(2) The intermediate "protocol layer" contains procedures *Handler*, *Pitcher*, and *Catcher*, that use the channel and give services to the users (which are not directly aware of these background procedures).

(3) The lower layer is the given raw channel. It is a no-loss, no-duplication channel in which disordering may occur. (A formal definition of these terms was given in Chapter 8, Section 1). Not just data messages from A to B, but acknowledgment messages from B to A can also arrive out of order.

From the point of view of the users, this is a producer/consumer protocol, and the correctness condition is that the nth *Receive* event obtains the value of the nth *Send* event, and that each such event is terminating.

Figure 9.42 depicts the five operation procedures in Stations A and B and their relationship with the channel and with the users' layer. The arrow from *Producer* to *Send* shows that *Producer* invokes procedure *Send* at Station A. Similarly, *Consumer* invokes *Receive* at Station B. Two instructions (external operations),

procedure *Producer*	procedure *Consumer*
begin	**begin**
repeat_forever	**repeat_forever**
(1) Produce_item (*a*);	(1) *Receive*(*b*);
(2) *Send*(*a*).	(2) Consume_item(*b*).
end.	**end.**

procedure *Handler*	procedure *Pitcher*	procedure *Catcher*
begin	**begin**	**begin**
repeat_forever	**repeat_forever**	**repeat_forever**
DataArr	*PitchAck*	*CatchAck*
end	**end**	**end**

protocol Sliding-Window
concurrently do *Producer, Consumer, Handler, Pitcher, Catcher* **od**

FIGURE 9.41. *Producer/Consumer* sending and receiving items.

ToChannel and *FromChannel*, are used by the operation processes to access the channel. Procedure *Send* invokes *ToChannel* to send its messages from Station A to B where it is received by *DataArr* in executions of *FromChannel*. Similarly, *PitchAck* invokes *ToChannel* at B, and *CatchAck* receives these messages at A.

(1) The messages going from A to B are of type *DataFrame* (see Figure 9). Each *DataFrame* d is a pair ($d.info, d.tag$) where the first field, $d.info$, is an *Item*, and the second field, $d.tag$, is a *TagNr*, that is a number in $\{0, \ldots, N-1\}$.
(2) The messages going from B to A are natural numbers, which we call *SequenceNr* in this context.

The aim of the protocols is to ensure that, despite the fact that the given raw channel may disorder messages, the *Send/Receive* operations form an orderly operating channel of capacity N (as described in Chapter 8 Section 1). We shall therefore concentrate on the *Send/Receive* procedures and prove that the higher level events they implement form indeed an orderly operating channel of capacity N. Roughly speaking this means that each sent item also arrives in order, and moreover *Producer* never has more than N outstanding (not acknowledged) items.

Figure 9.43 depicts the five operation procedures at the two stations, and their relationship with the channel and the different registers. Procedures' names are framed, and registers are underlined. The two arrows connecting *Receive* to *FrameExpected* indicate that procedure *Receive* both reads and writes on register *FrameExpected* (which carries sequence numbers). A single arrow from *FrameExpected* to *PitchAck* indicates that this procedure only reads register *FrameExpected*.

Also depicted in Figure 9.43 are the circular buffer-array *Buf* and the array of registers *Arrived*. Both are indexed from 0 to $N-1$. We assume safeness for each buffer cell $Buf[k]$ (for safeness see Chapter 6 Section 2). Observe that two

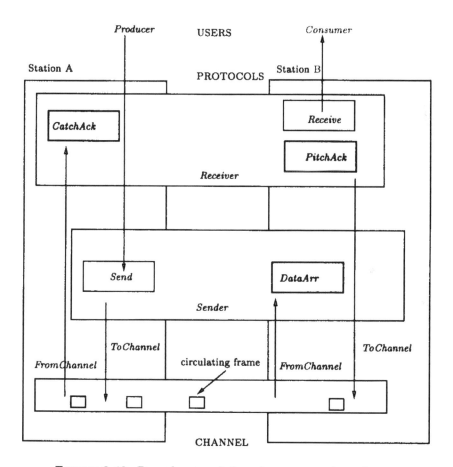

FIGURE 9.42. Procedures and three layers: raw channel, protocol, users.

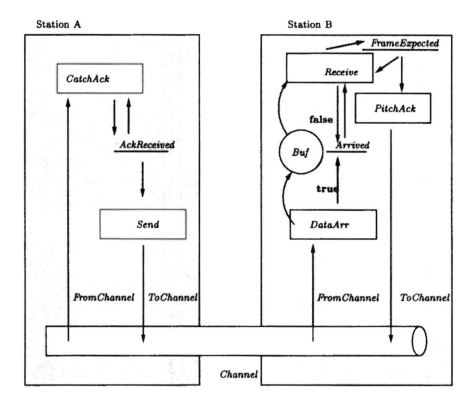

FIGURE 9.43. Procedures, channel, and shared variables.

> $SequenceNr = $ N;
> $TagNr = 0..N - 1$;
> $Item = $ data type of users' messages;
> $DataFrame = $ **record** $info$: $Item$;
> tag: $TagNr$;
>
> ---
>
> Local Registers: $AckReceived$, $Frame\_Expected$
> carry $SequenceNr$ and are initially 0.
> $Arrived[k]$ (for $0 \le k < N$) are boolean, initially
> $false$.
> Buffers: $Buf[k]$ for $0 \le k < N$ of type $Item$.

FIGURE 9.44. Types and initial values.

independent procedures $Receive$ and $DataArr$ can write on $Arrived$. We therefore assume that each register $Arrived[k]$ is serial, which means that some mutual exclusion mechanism ensures that read/write operations on this register never overlap.

Using a periodical clock or some other mechanism, $Receiver$ sends from time to time acknowledgments to $Sender$. For simplicity we do not describe this mechanism and just require that procedure $PitchAck$ is activated infinitely often in order to regulate the Producer's flow. These acknowledgments are sequence numbers; an acknowledgment with sequence number n indicates that Consumer has received all first n messages. The $Sender$ has "right" to send up to N messages without receiving any acknowledgment. This number N of possible "outstanding" messages is the number of buffer cells at the Consumer's side.

Procedure $Send$ (Figure 9.45) uses variable $next\_to\_send$ to keep the number of frames already sent. Thus $\lceil next\_to\_send \rceil_N$ is the tag of the frame to be sent. The initial value of $next\_to\_send$ is 0. So the first frame has tag 0, the second 1, the Nth 0 again, etc. Register $AckReceived$ keeps the number of frames for which acknowledgment has been received so far. The sender has a "window" of $N - (next\_to\_send - AckReceived)$ messages which it can send without waiting. $next\_to\_send$ and $AckReceived$ are variables of different nature: $next\_to\_send$ is a variable accessed only by procedure $Send$, but $AckReceived$ is a local register owned by $CatchAck$ that can also be read by $Send$ (see Chapter 5, Section 5 for definition of local registers). This means that procedure $CatchAck$ may treat $AckReceived$ as though it were a local variable, as in "**if** $AckReceived < ack$ **then** $AckReceived := ack$".

Similarly, $Frame\_Expected$ is another local register that is owned by $Receive$ which treats it as a local variable when it executes

$$Frame\_Expected := Frame\_Expected + 1.$$

Procedure $PitchAck$ can only read it, as it does in executing

$$\text{``}Read(Frame\_Expected, ack)\text{''}.$$

Procedure $Receive$ (Figure 9.45) uses this local register ($Frame\_Expected$) to hold the number of frames so far delivered to the $Consumer$ (or, equivalently, the number of times *Receive* was invoked). [*Frame_Expected*]$_N$ is thus the tag of the next frame to be delivered to the *Consumer*. An incoming frame with tag k is buffered in *Buf*[k] until all previous frames have already been delivered, and then it is ready to be delivered too.

The reader may want a general description of the proof and its definitions. The correctness of the protocol depends on quite a large set of assumptions concerning the diverse communication devices such as the channel and buffers. The channel, important as it is, is not the main ingredient in the proof, and it seems better to try to isolate it and to base the main lines of the proof on the buffers and the mechanism by which the processes regulate their messages, namely the sequence numbers and tag numbers. A major step in this modular isolation of the channel is our definition of a *SEND* event to comprise not only the *Send* operation execution (at Station *A*), but also the corresponding *DataArr* execution at Station *B*. The *SEND* event thus represents both the sending, reception, and loading in its buffer of a message—it crosses the channel, so to speak, and hides the details of the definition which rely on the assumed properties of the channel.

Even though the *Producer* is serial and the *Send* events are serially ordered, the *SEND* events may well overlap. The *RECEIVE* events are the executions of the *Receive* procedure at Station *B*. We shall also define a return function Γ associating with each *RECEIVE* event R the corresponding *SEND* event $\Gamma(R)$. A major feature of the protocol is the usage of acknowledgments to control the sending of messages. We shall define *ACK* events and a function, *activate*, from *SEND* events to *RECEIVE* events. *activate*(S) = R indicates that S, at Station *A*, "knows" about termination of *RECEIVE* event R and can base its subsequent sending events on this knowledge. The information about these higher-level events and functions that is relevant for the correctness proof is gathered in the Sliding Window Axioms. Once we prove that these axioms indeed hold in any execution, the conclusion that the protocol is correct is very short (Section 3).

1. Protocol analysis and definition of higher-level events

In this section we analyze the protocols, define higher-level events, and prove some simple properties.

We assume a non-duplicating, non-lossy channel, which may however disorder the messages. The formal specification of Chapter 8 Section 1 uses predicates *send/receive*, but we will use predicates *ToChannel/FromChannel* to denote channel's events. The function γ is defined on all terminating *FromChannel* events, with properties as described in that section.

We will prove here that the *Send/Receive* operations implement an order preserving channel of capacity N. In fact, after defining *SEND/RECEIVE/ACK* events we will prove that the Sliding Window Axioms of Chapter 8 hold in any execution of our protocol.

Let $\mathcal{H}$ be a two-level system execution of the protocol. That is, a system execution in which both the lower-level and the higher-level events are represented. The lower-level events include the reads and writes of the registers, the buffer load/unload events, the *ToChannel/FromChannel* lower-level events, and

---

procedure *Receive(b : Item)*
var x : *boolean*;
begin

(1) **repeat** *Read(Arrived[$\lceil FrameExpected \rceil_N$], x);*
    until $x =$ **true**;
(2) *unload(Buf[[$\lceil FrameExpected \rceil_N$], b);*
    {deliver to *Consumer* item b}
(3) *Write(Arrived[$\lceil Frame\_Expected \rceil_N$], false);*
(4) *FrameExpected:= FrameExpected + 1*

end

---

procedure *PitchAck*;
var *ack: SequenceNr*
begin
Read(Frame_Expected, ack);
ToChannel(ack)
end

---

procedure *CatchAck*;
var *ack : SequenceNr* ;
begin

(1) *FromChannel(ack);*
(2) **if** *AckReceived < ack*
        then *AckReceived := ack*

end

---

procedure *Send(a : Item)*;
var *s : DataFrame*;
next_to_send, ack : SequenceNr (both variables are initially 0);
begin

(1) **while** *next_to_send − ack = N*
    do *Read(AckReceived, ack);*
(2) *s.info := a* {item a is obtained
    from *Producer*}
    s.tag := $\lceil next\_to\_send \rceil_N$;
(3) *next_to_send := next_to_send + 1;*
(4) *ToChannel(s)*

end

---

procedure *DataArr*
var *r : DataFrame*
begin

(1) *FromChannel(r)*
(2) *load(Buf[r.tag], r.info);*
(3) *Write(Arrived[r.tag], true)*

end

FIGURE 9.45. The five operation procedures. *Receiver* comprises *Receive*, *PitchAck*, and *CatchAck*, all activated concurrently. *Frame_Expected* and *AckReceived* are initially 0. *Sender* includes *Send*, and *DataArr*.

the internal events, such as assignments to variables, or checking of conditions. These lower-level events are assembled to form higher-level events that represent executions of the procedures.

We are not interested in every lower-level event, and the proof is more natural if only the relevant communication events are included in our higher-level events. Thus, for any procedure P, we distinguish between an *execution* of P, which includes all the lower-level events as defined in Chapter 5, and the higher-level *event* corresponding to P which includes only those lower-level events that are introduced in the definitions below. For any procedure we adopt the procedure's name as a unary predicate that denotes $DataArr(X)$ these higher-level events.

(1) $DataArr(X)$ is defined when X is a higher-level event that comprises the following lower-level events corresponding to lines 1,2,3 in the procedure defined in Figure 9.45:

   (a) An execution of $FromChannel(r)$.
   If this execution e of $FromChannel(r)$ is not terminating, then X consists solely of e and is a non-terminating higher-level event, but otherwise X contains the following two additional events.

   (b) A load event in which item $r.info$ is loaded onto buffer cell $Buf[r.tag]$, where $r.tag$ is the tag value of r.

   (c) A write event: The value *true* is written onto register $Arrived[r.tag]$.

(2) $PitchAck(X)$ is defined to hold when X contains two events: A read of $Frame\_Expected$, and a sending of that number down the channel (from B to A). While the read of a register is always terminating, an execution of $ToChannel(ack)$ may be nonterminating, and in this case the $PitchAck$ event is nonterminating.

(3) $CatchAck(X)$ is defined in three cases as follows. Consider an execution of procedure $CatchAck$, and let f be the execution of line (1) (namely, "$FromChannel(ack)$").

   (a) If f is nonterminating, then form the higher-level event $X = \{f\}$ and define $CatchAck(X)$.

   (b) If f is terminating and condition $AckReceived < ack$ holds, then form $X = \{f, w\}$ where w is the write on $AckReceived$. Define $CatchAck(X)$. In this case X is said to be a successful $CatchAck$ event.

   (c) If f is terminating, but $AckReceived < ack$ does not hold in our execution, then form $X = \{f\}$, define $CatchAck(X)$, and say that X is a terminating non-successful $CatchAck$ event.

   So if $CatchAck(X)$, then X is terminating just in case the $FromChannel$ event in X is terminating, and in this case $ack(X)$ denotes the value of that $FromChannel$ event. $ack(X)$ is also the value of variable ack in this terminating execution.

An execution of a procedure in $\mathcal{H}$ that contain a **while** or **repeat-until** instruction is said to be *Successful* in case the execution of this loop is terminating. For example an execution of $Send$ is successful if the **while** loop of line 1 is terminating. (The procedure itself may be non terminating though, in case the execution of $ToChannel(s)$ is non terminating.)

So in any successful execution of *Send*,

$$(20) \qquad\qquad next\_to\_send - ack \neq N$$

holds at the exit from the **while** loop. Either because it holds at entry, and then the loop's body is not executed and the execution of the while instruction is said to be "immediate", or else because the last read of *AckReceived* in X, called "the successful read of X", determines a value of *ack* that satisfies condition (20). (The value of *ack* at the entry to X is determined by the previous execution of *Send*, or is the initial value 0 in case X is the first execution of *Send*.)

The higher-level *Send* events are defined by cases as follows. We define $Send(X)$ if X is a set of lower-level events of the following kind: Unsuccessful, Immediate, Not Immediate.

> **Unsuccessful:** X is an infinite collection of read events of register *Ack-Received*, corresponding to some execution of procedure *Send*, such that all of these events satisfy $next\_to\_send - ack = N$. Thus, in this case, X is a non-successful execution of procedure *Send*.
>
> **Immediate:** X corresponds to an immediate execution of the procedure. Then the body of the **while** loop is not executed and X contains a single event: the execution of $ToChannel(s)$ corresponding to line (4) of the protocol.
>
> **Not Immediate:** Then X consists of two events: The successful read of *AckReceived*, the read that finished the loop, and the sending of s to the channel.

If $Send(X)$ then X is terminating iff X is successful and its sending event is terminating.

For every higher-level *Send* event S we define $next\_to\_send(S)$ as the value of variable $next\_to\_send$ at entry to S. $next\_to\_send(S)$ is a *SequenceNr*.

On every *terminating* higher-level *Send* event S we define the following functions: $ack(S)$, $F\_Value(S)$, and $Content(S)$ as follows.

(1) $ack(S)$ is the value of variable *ack* exactly after the execution of the (terminating) while loop in S. So clearly

$$(21) \qquad\qquad next\_to\_send(S) - ack(S) \neq N$$

(whether S is immediate or not).

(2) $F\_Value(S)$ is the frame's value that S is sending. That is the value of variable s when line 4 is executed. If $v = F\_Value(S)$, then v is a pair $\langle v.tag, v.info \rangle$. Clearly, $F\_Value(S).tag = \lceil next\_to\_send(S) \rceil_N$, and $F\_Value(S).info$ is the value of variable a obtained from the *Producer* in line 2. We define $Content(S)$ to be that value of a. $Content(S)$ is an *Item*.

Let $\langle S_i \mid i \in I_0 \rangle$ be an enumeration of all *Send* events in $\mathcal{H}$ in increasing order. The index set I_0 may either be a finite initial interval of natural numbers or the set of all natural numbers. Increasing order means that $S_i \prec S_j$ holds in $\mathcal{H}$ whenever $i < j \in I_0$. So S_0 is the first *Send* event, S_1 the second, etc. Since we assume that *Producer* is active forever and keeps invoking *Send*, I_0 is finite exactly in case there is a non-terminating *Send* event, and in that case, if $i_0 = \max(I_0)$ then the (unique) non-terminating *Send* event is S_{i_0}. Thus $S_0, \ldots, S_{i-1}, S_{i_0}$ is an enumeration of *all Send* events if I_0 is finite, and $\langle S_i \mid i \in \mathbb{N} \rangle$ is an enumeration of all *Send* events if $I_0 = \mathbb{N}$. By definition, each terminating *Send* is also successful, but S_{i_0} which is not terminating may either be successful or not. If it is successful then the reason it is non terminating is that its *ToChannel* event is non terminating.

LEMMA 1.1. *For every* $i \in I_0$ *next_to_send*$(S_i) = i$.

Proof. The initial value of *next_to_send* is 0, and so *next_to_send*$(S_0) = 0$. Now variable *next_to_send* is increased by 1 in each execution of line (3) and hence *next_to_send*$(S_{i+1}) =$ *next_to_send*$(S_i) + 1$ follows. □

It follows from this lemma and formula (21) above that for every $i \in I_0$ if S_i is successful then

$$(22) \qquad\qquad i - ack(S_i) \neq N.$$

We now define higher-level *Receive* events:

DEFINITION 1.2 (RECEIVE EVENTS:). *Consider an execution of procedure Receive.*

(1) *If the execution is not successful, then the infinite set of all reads of* Arrived$[\lceil$ Frame_Expected$\rceil_N]$ *forms a higher-level event* X *and we define* Receive(X) *in this case.*

(2) *If the execution is successful, then* X *is formed as the collection containing*
   (a) *the successful read of* Arrived$[\lceil$ Frame_Expected$\rceil_N]$ *(that last read that returned the value true),*
   (b) *the unload event corresponding to line (2),*
   (c) *the write event corresponding to line (3), and*
   (d) *the increase event of* Frame_Expected.
   Clearly "terminating" and "successful" can be used interchangeably for Receive events.

We define the functions: *unload*(R), *Content*(R) and *Frame_Expected*(R) on terminating *Receive* events R as follows.

(1) The lower-level unload event in R corresponding to line 2 is denoted *unload*(R).
(2) *Content*(R) is the value of the unload operation done in R,

$$Content(R) = Value(unload(R)).$$

This is also the value of item b that is returned to the *Consumer*.

(3) The value of variable *Frame_Expected* at the entry to R is denoted *FrameExpected*(R). So *Frame_Expected*(R) is also defined when R is not terminating.

Let $\langle R_j \mid j \in J_0 \rangle$ be an enumeration of all *Receive* events in increasing order. $R_0 \prec R_1$ etc. The index set J_0 is an initial segment—finite or infinite—of the set of natural numbers. J_0 is finite exactly if some (unique) *Receive* event is not terminating (equivalently, not successful). In case J_0 is finite, we let $j_0 = \max(J_0)$. In this case

$$R_0, \ldots, R_{j_0}$$

is the sequence of *all Receive* events.

Since the initial value of *Frame_Expected* is 0, and as each terminating *Receive* increases the value of *Frame_Expected* by 1, we have:

LEMMA 1.3. $FrameExpected(R_j) = j$ *for every* $j \in J_0$.

So, if R_j is successful for $j \in J_0$, R_j unloads $Buf[\lceil j \rceil_N]$, and writes *false* on $Arrived[\lceil j \rceil_N]$.

The initial value *false* of registers $Arrived[k]$ (for $0 \le k < N$) is determined by some initial writes made by the system, but except for these initial writes, the only writes of value *false* on $Arrived[k]$ are those in successful *Receive* events $R = R_j$, where $Frame\_Expected(R) = j$ is such that $k = j \pmod{N}$.

We want to prove that every *Send* and *Receive* events are successful and terminating, but at this stage we do not even know that the successful *Send* events are terminating. Indeed, it is conceivable that a *Send* successfully terminates its while loop but never returns from executing its $ToChannel(s)$ instruction.

Recall (from Chapter 8, Section 1) that a channel is *attentive* at extremity $E = A, B$ if it either contains a non terminating *FromChannel* event, or else it contains an infinite number of *FromChannel* events. If the channel is attentive at extremity E_1, and E_2 is the other extremity, then the channel is non lossy exactly if the function γ defined on the terminating *FromChannel* events at E_1 is *onto* the set of *ToChannel* events at E_2.

LEMMA 1.4.   (1) *The channel is attentive at both Station A and Station B, and hence γ is onto the ToChannel events, and every ToChannel event is terminating. Therefore every successful Send operation is terminating.*
(2) *Every PitchAck event is terminating, and there are infinitely many such events. Hence there are infinitely many ToChannel events at Station B.*
(3) *Every FromChannel event executed at Station A is terminating. Hence all CatchAck events are terminating and there are infinitely many of them.*

Proof. To prove (1) notice that instructions *FromChannel* are executed at Station A by procedure *CatchAck* alone, and at Station B by *DataArr*. Both of these procedures are repeatedly executed and they contain no instructions other than *FromChannel* that could be non-terminating. Hence each station either contains infinitely many *FromChannel* events or else a non terminating *FromChannel* event. Thus, by definition, the channel is attentive, and as it is a no loss channel, every *ToChannel* is terminating and γ is onto the *ToChannel* events. That is, every *ToChannel* event T is of the form $T = \gamma(F)$ where F is at Station B if T is at A, and vice versa.

	Station A	Station B
ToChannel	Terminating	Terminating infinitely many
FromChannel	Terminating infinitely many	Section 3
PitchAck	******	Terminating infinitely many
Send	Terminating if successful	******
Receive	******	Terminating if successful
CatchAck	Terminating infinitely many	******
DataArr	******	Section 3

FIGURE 9.46. Summary of Lemma 1.4

From the fact deduced above that every *ToChannel* event at B is terminating it follows immediately that every *PitchAck* event is terminating. As *Receiver* executes *PitchAck* repeatedly there are infinitely many executions of this procedure. Hence (2).

Since we have proven that there are infinitely many *ToChannel* events at Station B, and that the function γ is taking the *FromChannel* events at A onto the *ToChannel* events at B, there must also be infinitely many *FromChannel* events at A. Hence each one is terminating, and this proves the third item. □

The results of our lemma are summarized in the chart of Figure 9.46. We have not proven yet that all the *FromChannel* events executed at Station B are terminating, nor that there are infinitely many *ToChannel* events at A. We shall do this later at Section 3 when proving the liveness property of the protocol.

LEMMA 1.5. *For every $0 \leq i < N$, $i \in I_0$, and S_i is immediate. $N \in I_0$, and S_N is not immediate.*

Proof. Initially, the value of variable *ack* in procedure *Send* is 0, and so for every $i < N$ expression *next_to_send − ack* is evaluated to i in S_i and the while loop is executed immediately (which means that its body is not activated and *ack* remains 0). Then the execution of "*ToChannel(s)*" must also terminate in S_i, by lemma 1.4(1).

Since S_{N-1} is immediate, it is terminating and thus S_N is defined ($N \in I_0$). So expression *next_to_send − ack* is evaluated to N at the entry to the while loop of S_N, and hence the loop is not immediate and its body is activated. If S_N is successful then this shows that S_N contains a successful read of *AckReceived*. □

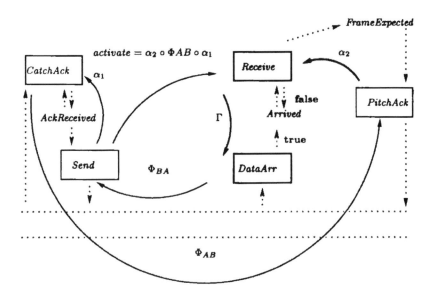

FIGURE 9.47. An illustration for definitions in subsection 1.1.

We need several more definitions of higher-level events and functions.

DEFINITION 1.6.   (1) *We have proved that every PitchAck event is terminating. If P is a PitchAck event, then define ack(P) to be the Sequence Number value of the read of register Frame_Expected in P, which is also the value of the ToChannel event in P, that is the value of the acknowledgment message.*

(2) *We have proved that all CatchAck events are terminating. Let C be a CatchAck event. The value of the FromChannel event in C is denoted ack(C). This is the value of variable ack in line (2). Not all CatchAck events necessarily write on register AckReceived. We have said that C is "successful" if condition AckReceived < ack holds, and then the assignment AckReceived := ack is executed in C.*

(3) *If H is a terminating DataArr event, then we define F_Value(H) as the value of the (terminating) FromChannel event in H (this data frame is also the resulting value of variable r).*

We also define load(H) to be the lower-level event in H that corresponds to the load instruction at line 2. It thus follows that if f = F_Value(H), then f.info = Value(load(H)) and the buffer cell loaded by load(H) is Buf[f.tag]. It is also clear that H contains a write of true on Arrived[f.tag].

1.1. Higher-level functions. We define six partial functions to serve us in the correctness proof. The definitions are illustrated in Figure 9.47

$$\alpha_1 \quad : \quad Send \to CatchAck$$

$$\alpha_2 \quad : \quad PitchAck \to Receive$$

$$\Phi_{BA} \quad : \quad DataArr \to Send$$

$$\Phi_{AB} \quad : \quad CatchAck \to PitchAck$$

$$activate \quad : \quad Send \to Receive$$

$$\Gamma \quad : \quad Receive \to DataArr$$

In the formulas above predicate names are used as domains and ranges of functions, that is as sets of higher-level events. So α_1, for example, is a partial function defined on *Send* events with value *CatchAck* events. We are going to define each of the functions and then state and prove some simple properties.

DEFINITION 1.7. *The function α_1 is defined over the terminating Send events with index $i \geq N$.*

Recall that $\langle S_i \mid i \in I_0 \rangle$ is the increasing enumeration of all Send events in $\mathcal{H}$. The function α_1 is defined now on all S_i's with $i \geq N$, by induction as follows. Suppose that $i \geq N$, $i \in I_0$, and S_i is terminating. There are two cases:

(1) *S_i is immediate: The successful wait-until loop is immediate and it needs no read. Then $i > N$ was proved in Lemma 1.5 (which showed that S_N is not immediate). Hence $i - 1 \geq N$, and $\alpha_1(S_{i-1})$ is defined. In this case we define*

$$\alpha_1(S_i) = \alpha_1(S_{i-1}).$$

(2) *S_i is not immediate: The successful wait-until loop of S_i is not immediate and it contains a successful read r of AckReceived. We will prove below that $\omega(r)$ is never the initial write on AckReceived, but meanwhile we cannot exclude this possibility. So, if $\omega(r)$ is the initial write on AckReceived than $\alpha_1(S_i)$ is defined to be that initial write, but if $\omega(r)$ is not the initial write on AckReceived, then $\omega(r)$ belongs to some unique successful CatchAck event C, and we define $\alpha_1(S_i) = C$.*

LEMMA 1.8. *Properties of α_1.*

(1) *Let S be some terminating S_i with $i \geq N$. If $C = \alpha_1(S)$ then $C \prec S$.*

(2) *The function α_1 is weakly monotonic on Send events: If $S_m \prec S_n$ are in the domain of α_1 (i.e., $m \geq N$) then $\alpha_1(S_m) \preceq \alpha_1(S_n)$.*

(3) *For every $i \in I_0$ with $i \geq N$, $\alpha_1(S_i)$ is a successful CatchAck event (i.e., $\alpha_1(S_i)$ is never the initial write on AckReceived). If $C = \alpha_1(S_i)$, then $ack(C) = ack(S_i)$.*

The proofs of 1 and 2 are by induction on n. To prove (1), suppose that $C = \alpha_1(S_i)$, where $i \geq N$ is in I_0.

There are two cases in the definition of $\alpha_1(S_i)$. In the first case we use the inductive assumption and obtain $C \prec S_{i-1} \prec S_i$. In the second case $\alpha_1(S_i)$ is defined by considering $\omega(r)$ where r is the successful read of *AckReceived* in S_i. Since r is the first event in S_i, and as (by seriality of the register) $\omega(r) \prec r$ and $\omega(r)$ is the last event in C, $C \prec S_i$ follows.

The proof of (2) is left to the reader.

We prove (3) by induction on $i \in I_0$. The base case is $i = N$. Then Formula 22 (on page 174) implies

$$N - ack(S_N) \neq N.$$

Hence, as acknowledgment numbers are ≥ 0, $ack(S_N) > 0$. But $ack(S)$ for a successful, non immediate *Send* is the value of the successful read in S. Thus $\omega(r)$ cannot be the initial write on *AckReceived* (which is a write of 0). So $\omega(r)$ belongs to some successful *CatchAck* event C, and $C = \alpha_1(S_N)$. $ack(C) = ack(S_N)$ now follows from the definitions (Definition 1.6 for $ack(C)$, and the definition on page 173 for $ack(S_N)$).

Now for $i > N$ induction and monotonicity imply that $\alpha_1(S_i)$ is not the initial write. Since only successful *CatchAck* events write on register *AckReceived*, the range of α_1 consists solely of successful *CatchAck* events. $\square$

Recall that every *PitchAck* event is terminating (by Lemma 1.4).

DEFINITION 1.9. *The function* α_2 : *PitchAck* $\rightarrow$ *Receive is defined on all PitchAck events. Let P be any PitchAck event, and let r be the read event of register Frame_Expected. Then $\omega(r)$ is the corresponding write on this (local) register. If $\omega(r)$ is the initial write (of value 0) on Frame_Expected, then $\alpha_2(P) = \omega(r)$. Otherwise, $\omega(r)$ belongs to some Receive event R, since only procedure Receive can write on Frame_Expected (in executing the instruction on line 4). We then define $R = \alpha_2(P)$. Since only successful Receive events contain a write on Frame_Expected, R is a successful Receive event.*

LEMMA 1.10. (1) *If $R = \alpha_2(P)$, then $R \prec P$, and if R is a Receive event then*

$$Frame\_Expected(R) + 1 = ack(P).$$

(2) α_2 *is weakly monotonic: If $P_1 \prec P_2$ are PitchAck events, then*

$$\alpha_2(P_1) \preceq \alpha_2(P_2).$$

(3) α_2 *is regular: If R is a Receive event such that $R \prec P$, then*

$$R \preceq \alpha_2(P).$$

Consequently, if $R_j \prec P$ then $j + 1 \leq ack(P)$. (Because $R_{j'} = \alpha_2(P)$ for some $j' \geq j$, and $j' + 1 = ack(P)$.)

Proof. Let P be some *PitchAck* event and let r be the read event in P. Then r is the first event in P. Assume that $R = \alpha_2(P)$ is a *Receive* event. By definition $\omega(r) \in R$ (or $\omega(r)$ is R if it is the initial write), and it is the last event in R. As $\omega(r) \prec r$, $R \prec P$ follows. By definition, *Frame_Expected*(R) is the value of that variable before the increase by 1, and hence the value of $\omega(r)$ is *Frame_Expected*$(R) + 1$. The first item of the lemma is thus concluded.

Next we prove weak monotonicity. Let $r_1 \in P_1$ and $r_2 \in P_2$ be the reads of register *Frame_Expected*. Then $P_1 \prec P_2$ implies $r_1 \prec r_2$ and hence $\omega(r_1) \preceq \omega(r_2)$ (this is part of the specification of ω). As *Receive* is a serial process, $\alpha_2(P_1) \preceq \alpha_2(P_2)$ follows from $\omega(r_i) \in \alpha_2(P_i)$. (If $\omega(r_2)$ is the initial write on *Frame_Expected*, then $\omega(r_1) = \omega(r_2)$, so that the weak monotonicity holds.)

The proof of (3) is left to the reader. $\square$

We define two functions. Φ_{BA} is defined on all terminating *DataArr* events, and Φ_{AB} is defined on all (terminating) *CatchAck* events.

DEFINITION 1.11.   (1) *For every terminating DataArr event H define a terminating Send event S, denoted $\Phi_{BA}(H)$, as follows. Let $f \in H$ be the execution of FromChannel(r) in H which is assumed to be terminating; then $\gamma(f)$ is the corresponding ToChannel event at Station A, and there is a single Send event in $\mathcal{H}$ containing $\gamma(f)$: we let $S = \Phi_{BA}(H)$ be this event. S is terminating because non terminating Send events do not contain any ToChannel event.*
As $\gamma(f)$ is the last lower-level event in $\Phi_{BA}(H)$, and as $\gamma(f) \prec f$,

$$S = \Phi_{BA}(H) \prec H$$

follows from the fact that f is the first event in H. Clearly

$$F\_Value(\Phi_{BA}(H)) = F\_Value(H).$$

We remark that Φ_{BA} is one-to-one (because γ is one-to-one, as the channel is non-duplicating). However Φ_{BA} need not be order preserving, since γ is not assumed to have this property.
(2) *To any CatchAck event C, the corresponding PitchAck event $P = \Phi_{AB}(C)$ is defined as follows. We know by Lemma 1.4 that every CatchAck event is terminating. Let $f \in C$ be the terminating FromChannel event. (As f is executed in Station A, $\gamma(f)$ is in Station B.) There is a unique PitchAck event P with $\gamma(f) \in P$. Then we define $P = \Phi_{AB}(C)$.*
It follows that $P \prec C$, since $\gamma(f) \prec f$. Clearly

$$ack(P) = ack(C)$$

in this case as $ack(C) = Value(f) = Value(\gamma(f)) - ack(P)$.
Φ_{AB} is clearly one-to-one since the channel is non-duplicating. Φ_{AB} is not necessarily monotonic, but its restriction to the successful events is, as we shall see in Lemma 1.14.

LEMMA 1.12.   (1) *Every terminating Send event S is of the form $\Phi_{BA}(H)$ for some (uniquely determined) terminating DataArr event H. Thus Φ_{BA} is a bijection (one-to-one and onto map) from the terminating DataArr events and onto the terminating Send events. We let Φ_{BA}^{-1} denotes its inverse.*
(2) *If $\Phi_{BA}(H) = S_i$, then H contains a load of $Buf[\lceil i \rceil_N]$ of value*

$$Content(S_i) = F\_Value(S_i).info,$$

and a write of value true on register $Arrived[\lceil i \rceil_N]$.
(3) *There is a non terminating Send event iff there is a non terminating DataArr event.*

Proof. Let S be a terminating *Send* event, and let $t \in S$ be the *ToChannel* event obtained in executing line 4. We first remark that t is in the range of γ. This follows from Lemma 1.4(1), saying that Station B of the channel is attentive and hence that every *ToChannel* event executed at A is terminating and in the range of γ (since the channel is non lossy).

So let f be the unique *FromChannel* event with $\gamma(f) = t$, and let H be the unique *DataArr* event containing f. Then $\Phi_{BA}(H) = S$ by definition. f is terminating because the domain of γ consists only of terminating *FromChannel* events, and hence H is terminating as well.

We now prove the second item. Suppose that $S_i = \Phi_{BA}(H)$ is the ith terminating *Send* event. We know (Lemma 1.1) that $i = next\_to\_send(S_i)$. Then $\lceil i \rceil_N$ is the tag field of the *DataFrame* message sent to the channel in executing line (4) in S_i. But $\Phi_{BA}(H) = S_i$ implies that $F\_Value(S_i) = F\_Value(H)$, that is that the value of the frame sent by S_i equals that of the frame received by H. Thus the value of $r.tag$ in H is $\lceil i \rceil_N$ which implies the result

The last claim of the lemma clearly follows from (1) of our lemma, which implies that the number of terminating *Send* events is finite iff the number of terminating *DataArr* events is finite. There exists a non-terminating *Send* event (*DataArr* event) iff the number of terminating *Send* events (*DataArr* events respectively) is finite. Hence (3). $\square$

The enumeration $\langle S_i \mid i \in I_0 \rangle$ of all *Send* events in $\mathcal{H}$ induces an enumeration $\langle H_i | i \in I_0 \rangle$ of all terminating *DataArr* events. If $I_0 = \mathbb{N}$ then $\cdot H_i = \Phi_{BA}^{-1}(S_i)$. If I_0 is finite and S_{i_0} is the non terminating *Send* event, then there is a non terminating *DataArr* event which we denote by H_{i_0}. We remind the reader that Φ_{BA} is not necessarily order preserving and therefore $\langle H_i \mid i \in I_0 \rangle$ is not an increasing enumeration.

We now look at the *Receive* events. Recall that $\langle R_j | j \in J_0 \rangle$ is an enumeration of all *Receive* events in order.

LEMMA 1.13. *Suppose that J_0 is finite and let R_{j_0} be the last (non terminating) Receive event. If $j_0 \in I_0$ and H_{j_0} is terminating, then $j_0 \geq N$ and the situation*

$$(23) \qquad\qquad R_{j_0 - N} \prec H_{j_0}$$

is impossible.

Proof. The elegant way of proving this is by reference to the *KanGaroo/LoGaroo* protocol and to Theorem 3.1 in Chapter 5. Procedure *DataArr* is *Kan*, with "*FromChannel*" being "obtain_a_tag", and *Receive* is *Lo*. (Buffer operations disregarded.) Theorem 3.1 then gives our lemma, since $tag(H_{j_0}) = j_0$. However, we repeat the argument for the reader's convenience.

Suppose on the contrary that R_{j_0} is not terminating but H_{j_0} is terminating. Since R_{j_0} is not terminating it contains infinitely many unsuccessful reads of *Arrived*$[[j_0]]$, all returning *false*. Let r be one of those reads for which $H_{j_0} \prec r$ holds. Since H_{j_0} contains a write of *true* on *Arrived*$[\lceil j_0 \rceil_N]$, there is a write of *false* on this register in between $H_{j_0 - N}$ and r, namely $\omega(r)$. As $H_{j_0} \prec \omega(r)$, $\omega(r)$ is not the initial write and hence it is in some terminating *Receive* event R_j with $\lceil j \rceil_N = \lceil j_0 \rceil_n$ (as only these *Receive* events contain a write of *false* on *Arrived*$[\lceil j \rceil_N]$). As R_j is terminating, $j < j_0$, and thus $j \leq j_0 - N$. Since $\omega(r) \in R_j \preceq R_{j_0 - N}$ and $H_{j_0} \prec \omega(r)$, $R_{j_0 - N} \prec H_{j_0}$ is impossible. $\square$

LEMMA 1.14. *The function* $\Phi_{AB} : CatchAck \rightarrow PitchAck$ *is onto the PitchAck events, and is monotonic on the successful CatchAck events.*

Proof. That every *PitchAck* event P is of the form $\Phi_{AB}(C)$ for some *CatchAck* event C is a consequence of Lemma 1.4, which says that γ is onto the set of *ToChannel* events at Station B.

Recall that C is a successful *CatchAck* event if condition "*AckReceived* < *ack*" holds. As the channel is not assumed to be order preserving, Φ_{AB} is not necessarily monotonic outside the successful events. However monotonicity will be proved on the successful *CatchAck* events. Since the value of register *Frame_Expected* can only increase by *Receive*, if $r_1 \prec r_2$ are two reads of *Frame_Expected* then $Value(r_1) \leq Value(r_2)$. Hence if $P_1 \prec P_2$ are two *PitchAck* events then $ack(P_1) \leq ack(P_2)$. Now if $C_1 \prec C_2$ are two *CatchAck* events and C_2 is successful, then $ack(C_1) < ack(C_2)$, because "*AckReceived* < *ack*" holds in C_2. Hence if $P_1 = \Phi_{AB}(C_1)$, $P_2 = \Phi_{AB}(C_2)$, and $C_1 \prec C_2$, then $P_1 \prec P_2$ (as $ack(P_1) < ack(P_2)$ follows).

This proof omits some details, but the reader has surely noticed the similarity between the *PitchAck/CatchAck* procedures and the *PITCH/CATCH* procedures of Chapter 5. The detailed proof given there to Lemma 1.6, showing the monotonicity of Φ on the successful *CATCH* events, can be seen as a formal proof of our lemma. $\square$

DEFINITION 1.15. *Let activate be the composition* $\alpha_2 \circ \Phi_{AB} \circ \alpha_1$. *That is, for every successful Send event* S_i *with* $i \geq N$,

$$activate(S_i) = \alpha_2(\Phi_{AB}(\alpha_1(S_i))).$$

By Lemma 1.8(3), $\alpha_1(S_i)$ is a successful *CatchAck* event C, and hence $ack(C) > 0$. Hence, if $P = \Phi_{AB}(C)$, then $ack(P) > 0$. Hence $R_j = \alpha_2(P)$ is not the initial write on *Frame_Expected*, but rather a successful *Receive* event. So $activate(S_i) = R_j$ for some $j \in J_0$.

LEMMA 1.16. (1) *If* $S = S_i$ *is a successful Send event with* $i \geq N$ *and* $R_j = activate(S)$, *then* $j + 1 = ack(S)$, *and therefore* $i - (j + 1) \neq N$.

(2) *If* $S = S_i$ *is a successful Send event with* $i \geq N$ *and* $R = activate(S)$, *then* $R \prec S$.

(3) *activate is weakly monotonic: If* $i_2 \geq i_1 \geq N$ *are in* I_0, *then*

$$activate(S_{i_1}) \preceq activate(S_{i_2}).$$

(4) *If* $R_j = activate(S_i)$ *then* $i \leq j + N$.

Proof. We follow the definition of $activate(S)$ for $S = S_i$: Let $C = \alpha_1(S)$. Then C is a successful *CatchAck* event. Let $P = \Phi_{AB}(C)$ be the corresponding *PitchAck* event, and then $R_j = \alpha_2(P)$, by definition of *activate*. We have the following equations which prove $j + 1 = ack(S)$.

$$
\begin{aligned}
ack(S) &= ack(\alpha_1(S)) \text{ (Lemma 1.8(3))} \\
ack(C) &= ack(\Phi_{AB}(C)) \text{ (observed in Definition 1.11(2))} \\
ack(P) &= Frame\_Expected(\alpha_2(P)) + 1 \text{ (by Lemma 1.10(1))} \\
Frame\_Expected(R_j) &= j \text{ (by Lemma 1.3)}
\end{aligned}
$$

Now Equation (22) on page 174 gives $i - ack(S_i) \neq N$, and hence $i - (j+1) \neq N$.

Next we prove the second item. By Lemma 1.8(1), $C = \alpha_1(S) \prec S$. When Φ_{AB} was defined we remarked that $P = \Phi_{AB}(C) \prec C$, and in Lemma 1.10(1) we proved $R = \alpha_2(P) \prec P$. Since $activate$ is the composition of these maps, $R \prec S$.

We now prove weak monotonicity of $activate$. Lemma 1.8(2)(3) says that α_1 is weakly monotonic, and that the range of α_1 contains only successful $CatchAck$ events. Φ_{AB} is monotonic on these events (Lemma 1.14), and α_2 is weakly monotonic (Lemma 1.10)(2). Hence the composition $activate$ is weakly monotonic.

The proof of (4) is by induction on i, for $i \geq N$. Case $i = N$ is trivial, as $j \geq 0$. Suppose that $R_j = activate(S_i)$, but $N + j < i$. So induction can be applied to $N + j$: Let $R_{j'} = activate(S_{N+j})$, and thus $(N + j) \leq j' + N$. So

$$
j \leq j'.
$$

Yet, by weak monotonicity of $activate$, and since $N + j < i$, $j' \leq j$ follows. Hence $j' = j$. By monotonicity again, applied to $N + j < N + j + 1 \leq i$, $R_j = activate(S_{N+j+1})$. This is impossible by item (1) of the lemma. $\square$

Now the function Γ is defined.

DEFINITION 1.17 $(\Gamma(R))$. *Let R be one of the successful Receive events in $\mathcal{H}$. Look at the execution of line 1:*

$$
\textbf{repeat } Read(Arrived[\lceil FrameExpected \rceil_N], x) \textbf{ until } x = true;
$$

Since R is successful, the semantics of this loop implies that it contains a successful read of $Arrived[\lceil FrameExpected \rceil_N]$ which returned the value true. Let $r \in R$ be the successful read, and $\omega(r)$ be the corresponding write. Since only terminating DataArr contains the instruction to write true on Arrived (and as the initial value of this register is false), there is a unique DataArr event H with $\omega(r) \in H$, and we define

$$
\Gamma(R) = H.
$$

As r is the first event in R and $\omega(r)$ is the last event in H, $H \prec R$ follows.

LEMMA 1.18. (1) Γ *is one-to-one.*
(2) *If $H = \Gamma(R)$ then $H \prec R$. Moreover, if $k = \lceil Frame\_Expected(R) \rceil_N$ then $F\_Value(H).tag = k$; so that both $load(H)$ and $unload(R)$ are operations on $Buf[k]$.*

Proof. That Γ is one-to-one was concluded already in a different setting when the KanGaroo LoGaroo protocol was investigated, and the proof was presented in Chapter 5 Theorem 3.2. So we only outline the argument. Take two successful *Receive* events R_1 and R_2. By seriality of the *Receive* events we may assume that $R_1 \prec R_2$, and let r_1 and r_2 be the successful reads of *Arrived*[k_1] and *Arrived*[k_2] in R_1 and R_2 respectively. If $k_1 \neq k_2$ then clearly $\omega(r_1) \neq \omega(r_2)$ and then $\Gamma(R_1) \neq \Gamma(R_2)$. If $k_1 = k_2$ (=k say) then R_1 contains a write of value *false* on *Arrived*[k]. It follows that $R_1 \prec \omega(r_2)$ and hence that $\omega(r_1) \prec \omega(r_1)$.

We now prove the second item of the lemma. In defining Γ we have noted that $\Gamma(R) \prec R$. Recall that *Frame_Expected*(R) is the value of that register in executing line (1), and that $F\_Value(H)$ is the value of the frame obtained by H from the channel. Assume $k = \lceil Frame\_Expected(R)\rceil_N$, and $k' = F\_Value(H).tag$. We must prove $k = k'$. Let $r \in R$ be the successful read in R, and $\omega(r)$ be the corresponding write. Then r is a read of *Arrived*[k]. but $\omega(r)$ belongs to H by definition of $H = \Gamma(R)$, and is the execution of line (3) in $\mathcal{H}$ (the instruction "$WRITE(Arrived[r.tag], \mathbf{true})$" in $\mathcal{H}$. Hence $\omega(r)$ is a write on *Arrived*[k']. Thus $k = k'$, because ω takes reads of a register and returns writes on that same register. By definition, *load*(H) is the lower-level load event in H. It follows that *load*(H) and *unload*(R) are operations on *Buf*[k]. $\square$

2. The Sliding Window Axioms hold

The reader who has reached this point may want to have some plan of the proof. We deal here with quite a complex protocol comprised of five concurrently operating procedures, and we need therefore some guiding ideas to organize the proof. The focal device of the protocol is the circular buffer (not the channel). The protocol aims to deposit the messages in this buffer, and to withdraw them in order afterwards. The channel is just a means of transporting the messages nearer to the buffer, and its presence only obscures the functioning of the protocol. If so, the proof should be so organized that the channel is hidden. For that reason we form complete events: A complete event consists of both the sending of a message and its receipt at the other extremity of the channel. By considering complete events and their properties, the channel dissolves and the main points of the protocol transpire: namely the higher-level Sliding-Window axioms.

2.1. Complete *SEND*/ *RECEIVE*/ *ACK* events. The following higher-level events are defined next:

(1) *SEND* events, which are obtained as unions of *Send* events and their corresponding *DataArr* events.
(2) *RECEIVE* events.
(3) Complete acknowledgment events, denoted *ACK*, formed by unifying *PitchAck* and *CatchAck* events.

(Recall that in order to have the finiteness condition, the higher-level events defined must be pairwise disjoint.)

DEFINITION 2.1.   (1) *We know from Lemma 1.12 that Φ_{BA} is a bijection from the terminating DataArr events onto the terminating Send events.*

The pairs $\langle \Phi_{BA}(H), H \rangle$ *are "glued" together in the following definition to form higher-level events called SEND events.*

Let H be a terminating DataArr event and $S = \Phi_{BA}(H)$ *be the corresponding terminating Send event. Then* $E = S \cup H$ *is called a SEND event (the union of two higher-level events is obtained, of course, by considering them as sets of lower-level events). S is called the Send part of E, and H is the DataArr part. E is a terminating higher-level event since both S and H are.*

If there is a nonterminating Send event (we will prove that there is none) then it is said to be a non-terminating SEND event.

Recall that $\langle S_i \mid i \in I_0 \rangle$ *is an enumeration of all Send events in* $\mathcal{H}$. *Since* Φ_{BA} *is a bijection this enumeration induces an enumeration of all terminating DataArr events: If* S_i *is terminating, we have defined* H_i *as that unique DataArr event with* $\Phi_{BA}(H_i) = S_i$, *and we set* $E_i = S_i \cup H_i$. *If* I_0 *is finite and* S_{i_0} *non terminating, then* $E_{i_0} = S_{i_0}$ *is the non terminating SEND event. So* $\langle E_i \mid i \in I_0 \rangle$ *is an enumeration of all SEND events. (Observe that the SEND events are not necessarily serially ordered.)*

(2) *The RECEIVE events are just the Receive events. So* $\langle R_j \mid j \in J_0 \rangle$ *is an enumeration of all RECEIVE events.*

(3) *For any terminating CatchAck event C, let* $P = \Phi_{AB}(C)$ *be the corresponding PitchAck event. Then* $ack(P) = ack(C)$ *(we noted this in Definition 1.11). In this case we form*

$$A = C \cup P$$

as an ACK event. We define then $ack(A) = ack(C) = ack(P)$. *Since* Φ_{AB} *is one-to-one, the ACK events are pairwise disjoint, and as there are infinitely many PitchAck events there are infinitely many ACK events.*

Our aim is to prove the Sliding Window Axioms (Chapter 8, Figure 8.38) for $\langle R_j \mid j \in J_0 \rangle$, $\langle E_i \mid i \in I_0 \rangle$ (instead of the S_i series) and the functions *activate* and Γ defined below.

The function *activate* : *Send* $\rightarrow$ *Receive* (Definition 1.15) can be extended naturally to a function *activate* : *SEND* $\rightarrow$ *RECEIVE*, defined on the (terminating) *SEND* events E_i with $i \geq N$ by $activate(E_i) = activate(S_i)$.

The function Γ : *Receive* $\rightarrow$ *DataArr* will be redefined as Γ : *RECEIVE* $\rightarrow$ *SEND* as follows. The function Γ is defined on the terminating *RECEIVE* events: If R_j terminates then $H = \Gamma(R_j)$ is a terminating *DataArr* event. Then $S = \Phi_{BA}(H)$ is defined and $E = S \cup H$ is a terminating *SEND* event. We define $E = \Gamma(R_j)$.

The first two axioms are obvious. In particular $E_i \prec E_j$ implies $i < j$, because $E_i \prec E_j$ implies that the *Send* part of E_i precedes the *Send* part of E_j.

The following two properties are now easily obtained from Lemma 1.16.

(24)   If $R = activate(E)$, then $R \prec E$.

This follows from Lemma 1.16(2) and from the fact that if $E = S \cup H$ where $S = \Phi(H)$, then $S \prec H$.

(25)   If $R_j = activate(E_i)$, then $i \leq j + N$.

This proves Axiom 3.

The following properties of Γ can be obtained from our results; they show that Γ satisfies Axiom 4.

(a) If $E_i = \Gamma(R_j)$ then
$$E_i \prec R_j.$$

(b) Γ is one-to-one on the terminating *RECEIVE* events. This follows from Lemma 1.18.

(c) If $E_i = \Gamma(R_j)$ then $i = j \pmod{N}$. This is a consequence of the following equalities.

$$
\begin{aligned}
Frame\_Expected(R_j) &= j \text{ (Lemma 1.3)} \\
\lceil j \rceil_N &= F\_Value(H).tag \text{ (by 1.18(2))} \\
F\_Value(H) &= F\_Value(S_i) \\
F\_Value(S_i).tag &= \lceil next\_to\_send(S_i) \rceil_N \\
&= \lceil i \rceil_N \text{ (Lemma 1.1)}.
\end{aligned}
$$

Axiom 5(a) is a consequence of the following lemma.

LEMMA 2.2. *If A is an ACK event, and $R_j \prec A$ is a Receive event, then $j + 1 \leq ack(A)$.*

Proof. Suppose that $A = C \cup P$ is an *ACK* event. Then $ack(A) = ack(P)$ by definition. If $R_j \prec A$ then $R_j \prec P$ and Lemma 1.10(3) implies that $j + 1 \leq ack(P)$. $\square$

Now if R_j is any terminating *RECEIVE* event, then there is an *ACK* event A such that $R_j \prec A$ (since there are infinitely many *ACK* events and as the finiteness property holds). Thus 5(a) follows.

Axiom 5(b) is obtained from the following lemma.

LEMMA 2.3. *Suppose that I_0 is finite and $i_0 = \max(I_0)$. Then $i_0 \geq N$ and there are in the system execution only ACK events A with $ack(A) \leq i - N$.*

Proof. We have proved in Lemma 1.5 that for every $\ell < N$, $\ell \in I_0$ and S_ℓ is successful (and is hence terminating). Since S_{i_0} is not terminating, $i_0 \geq N$. We have proved (in Lemma 1.1) that $next\_to\_send(S_{i_0}) = i_0$.

Since S_0 is non-terminating, it is not successful and it consists of an infinite set of read events of *AckReceived* which all returned $Value(r)$ such that $i_0 - Value(r) = N$ holds (any other value would lead to a successful read). So $Value(r) = i_0 - N$.

If there were a single *CatchAck* event C with $ack(C) > i_o - N$, then any of the read events r in S_{i_0} with $C \prec r$ would obtain $Value(r) \geq ack(C) > i_0 - N$, which brings a contradiction. $\square$

Now only Axiom 6 remains. It is essentially a restatement of Lemma 1.13. That is, while that lemma concludes $\neg(R_{j_0-N} \prec H_{j_0})$, we want $\neg(R_{j_0-N} \prec E_{j_0})$.

But since H_{j_0} is a subset of E_{j_0}, the desired conclusion follows, and the last of the Sliding Window Axioms is proved to hold.

3. Correctness of the protocol

Recall our aim: to prove that the *Send/Receive* events form a FIFO channel of capacity N. Since the Sliding Window Axioms hold, so do their consequences from Chapter 8:

$$I_0 = J_0 = \mathbf{N},$$

$$\Gamma(R_k) = E_k,$$

and for every $0 \leq k < N$ the events $\{E_i \mid i \in I_0 \wedge i = k \pmod{N}\}$ and $\{R_j \mid j \in J_0 \wedge j = k \pmod{N}\}$ are conflict free (see Lemma 3.4 there). Moreover, the function *activate* satisfies the properties established in Lemma 3.5 in Chapter 8. Namely that if $R_j = activate(S_i)$, then $j + 1 \leq i \leq j + N$.

Recall (Definition 2.1) that $E_i = S_i \cup H_i$ for $i \in I_0$, and $\Phi_{BA}(H_i) = S_i$. We also noticed there that $F\_Value(S_i) = F\_Value(H_i)$ (the frame received by H_i was sent by S_i). In Lemma 1.12 we proved that H_i contains a load of $Buf[\lceil i \rceil_N]$ of value $F\_Value(S_i).info = Content(S_i)$.

Now, since H_i is a subset of E_i, the events $\{H_i \mid i = k \pmod{N}\}$ and $\{R_j \mid j = k \pmod{N}\}$ are conflict free (for $0 \leq k < N$). The safeness of $Buf[k]$ implies that for every n the value of the nth unload of the buffer is the value of the nth load, and so

$$Content(R_i) = Content(S_i)$$

follows for all $i \in I_0$. Since $I_0 = J_0 = \mathbf{N}$, all S_i and R_i are terminating. This proves that the *Send/Receive* events implement a FIFO channel. In addition, the conflict freedom includes the fact that

$$R_j \prec S_{j+N}$$

which implies that when the $(j+N)$th *Send* operation S_{j+N} has terminated, the sender knows that the message of S_j has already been received. This was defined as the "capacity N" property of the channel (when all events are terminating— see Equation 16 on page 152).

$DataArr(X)$ 172	$PitchAck(X)$ 207
$CatchAck(X)$ 172	$Successful$ 172
$Immediate$ 173	$Send(X)$ 173
$next\_to\_send(S)$ 173	$ack(S)$ 173
$F\_Value(S)$ 173	$Content(S)$ 173
$\langle S_i \mid i \in I_0 \rangle$ 173	$Receive(X)$ 174
$unload(R)$ 174	$Content(R)$ 174
$Frame\_Expected(R)$ 174	$\langle R_j \mid j \in J_0 \rangle$ 175
$ack(P), ack(C)$ 177	$F\_Value(H)$ 177
$load(H)$ 177	α_1 178
α_2 179	Φ_{BA} 180
Φ_{AB} 180	$activate$ 182
Γ 183	Complete events 184
$SEND$ 184	E_i 185
$RECEIVE$ 185	ACK 185

TABLE 9.2. Chapter's index.

10
Broadcasting and causal ordering

*In which a language for specifying networks is described, and some
important concepts are defined and discussed: causality, delivery,
and time-stamps.*

A network is shared by several processes that exchange information by sending
and receiving messages. A sending process can broadcast a message to all other
processes in a single operation. A distinction is made between broadcasting and
multicasting. In a broadcast the message is sent to all the other processes, but
in a multicast the sender can specify a group address to which the message is
destined. For simplicity we deal here only with broadcasting.

Group communication is an active field of research and the basic notions are
still taking shape as the investigators and builders of communication nets try to
find out which concepts are more useful. In writing this chapter I was mostly
influenced by the works of the Isis and the Transis teams in formulating the
question and specifying the requirements from the solution. Our setting here is
simpler because we assume that there are no partitions. (For just a glimpse of
this subject see [14], [10].)

There is a great diversity in the possible choice of signatures and first-order
properties that can be used to model networks. One of the decisions to make
is to determine which events are terminating. For simplicity, we assume here
that all send events in a network are terminating, while receive events may be
non-terminating. This is in contrast to the channels specified in Chapter 8 in
which both send and receive events could be non-terminating.

We shall describe two settings for specifying networks, that is, two signatures
and first-order languages in which to express properties of network activity:
(1) The Send/Receive Network Signature, and (2) the Message Domain Signature. In the following chapter we employ both signatures in one protocol: the
Send/receive signature is used to describe the given, raw net, and the Message
Domain is used to describe higher constructs implemented by the protocol.

1. *Send/Receive* Network Signature

We model the activity of a network (called N) with system executions for the following signature.

There are three sorts: *Processes, N_Messages,* and *Events.*

(1) *Processes*: This sort contains the constants $p_1, \ldots, p_{k_0}$. We think of p_i as a (name of a) machine or a process (usually serial) and for simplicity we may identify p_i with the number i. In the first part of the book we viewed a process as a predicate defined on the events, but here a process is a constant, a name, and the function *ProcId* gives for every event its owner—the process to which the event belongs.

(2) *N_Messages*: This sort contains the values that are sent and received by the network.

(3) *Events*: the events can be terminating or non-terminating.

The signature also contains the following predicates and function symbols.

(1) *N_SendEvent* and *N_RecEvent* are unary predicates defined on the events. The letter N refers to the name of the network. (In case there are different networks one can obtain greater clarity by using different sets of predicates.) The intention is to use $N\_SendEvent(e)$ and $N\_RecEvent(e')$ to express the fact that e and e' are a send and receive events respectively.

(2) *Content* is a function that assigns to each send event and to each terminating receive event its value in *N_Messages*. We express this by writing

$$Content : N\_SendEvent \cup (Terminating \cap N\_RecEvent) \rightarrow N\_Messages.$$

In this formula the predicates are used to represent sets of events.

(3) *ProcId* : *Event* $\rightarrow$ *Processes* is the function that assigns to each event its "owner". If $ProcId(e) = p$, then we say informally that e is executed by p, or that e belongs to p.

(4) $\gamma : (Terminating \cap N\_RecEvent) \rightarrow N\_SendEvent$ is the "return" function. It assigns to each terminating receive event its corresponding send event. If $\gamma(r) = s$ then we say that r obtained the message sent by s.

This ends the description of the *Send/Receive* Network Signature. The following properties (axioms) are always assumed to hold in any network.

Send/Receive Axioms

(1) All *N_SendEvents* are terminating.
(2) If $N\_RecEvent(r)$ and $Terminating(r)$, then
   $Content(r) = Content(\gamma(r))$, and $\gamma(r) \prec r$.

We shall make now some important definitions of properties that a *Send/Receive* network may have in a system execution.

DEFINITION 1.1. *Let S be a system execution in this Send/Receive network signature.*

(1) *We say that the network N is non-duplicating if γ is one-to-one when restricted to any one process. (Of course $\gamma(r_1) = \gamma(r_2)$ is possible if $ProcId(r_1) \neq ProcId(r_2)$.)*

(2) *A process q is said to be* attentive *in S if either there is a non-terminating N_RecEvent e with ProcId(e) = q, or else there are infinitely many N_RecEvents e with ProcId(e) = q. We say that the network is attentive in S iff every process is attentive.*

(3) *We say that the network is* non lossy *in S if for every attentive process q every N_SendEvent s in S is of the form s = γ(r) where r is a terminating N_RecEvent in q.*

(4) *We say that an attentive network is* FIFO *if for every two processes p and q the events pertaining only to the message flow from p to q form an orderly operating channel. More precisely, the function γ restricted to those N_RecEvents r in q with γ(r) in p is order preserving and onto the N_SendEvents in p.*

Some remarks are needed to clarify these definitions. A process is not attentive in S if it has only a finite number of *N_RecEvents*, and they all terminate. Presumably the program of that process executes only k *N_RecEvents*, and then does something else, never executing *Receive* again. In that case only k messages were received, and other messages are lost, but one does not blame the network for this, since the process is not attentive.

We have defined a no-loss network as one in which every message sent is received by every attentive process. A non attentive process with k receive events may receive only k messages, of course, even in a no-loss network.

It is easy to prove that if a system execution models an attentive FIFO network, then that network is non lossy.

2. Message Domain

The network signature defined in the previous section does not mention messages at all—only sending and receiving events together with their values are represented in the model. Many writers prefer to have messages explicitly in their models, as those objects that are being sent and received. In this section we extend the network signature and represent messages directly. The resulting signature is called the Message Domain Signature. We begin with some motivating remarks.

Each message is issued (sent) by some process p, called the sender of that message. The messages sent by any process p are enumerated $m_0^p, m_1^p, \ldots$, where m_0^p is the first message, m_1^p the second etc. Formally, the messages are pairs of the form $m_i^p = \langle p, i \rangle$ where process p is the sender of the message and i is its message number. The function *Content* gives to each message m its *Data* value *Content(m)*. It is obviously possible for two messages to have the same content.

In order to relate the messages to the send/receive events, two functions are needed: *SendMsg* and *RecMsg*. For every message m, *SendMsg(m)* is the event of sending m. Corresponding to our simplifying assumption that all send events are terminating, we assume here that for every message m, *SendMsg(m)* is terminating. In the applications given here, each process issues infinitely many messages, and therefore every pair $\langle p, i \rangle$ represents a message and *SendMsg(p, i)* is defined. If only a finite number of messages is sent, then *SendMsg* can be made a partial function, defined only on those messages that were actually sent.

In case m is received by process q, then $RecMsg(m, q)$ is that receive event. So, $RecMsg$ is a partial binary function—partial since it may well be the case that a message is not received by all processes. When it is defined, $RecMsg(m, q)$ is a terminating event. Non terminating receive events are possible, but they are not related to any message. If the network allows duplication of messages then $RecMsg(m, q)$ may denote only one of the receive events of m by P_q, and it is natural to require that $RecMsg(m, q)$ denotes the first receive event by q of message m. Thus $RecMsg$ may not be onto the receive events at P_q if duplication is possible.

It is also convenient to have a function that assigns to each event e, whether a send or a receipt of a message, that message with which e is concerned. If e is a terminating send/receive event of message m, then $message(e) = m$. If duplication of messages is impossible, then $message$ defined on the receive events by P_q is one-to-one.

We shall list now the sorts, predicates, function symbols, and constants that comprise the Message Domain Signature.

Sorts: There are four sorts.

> **Processes:** This sort contains a finite list $p_1, \ldots, p_{k_0}$ of "processes" (we shall identify p_j with j).
>
> **Data:** This is the useful data which is the content of the messages (useful, that is, from the point of view of the users of the system).
>
> **MessageNumbers:** This is the set of the natural numbers with their ordering $<$. It is used to identify messages issued by each process.
>
> **Messages:** A message is a pair $m = \langle p, i \rangle$ where p is in *Processes* and i is in *MessageNumbers*. So
>
> $$Messages = Processes \times MessageNumbers.$$
>
> We also write $m = m_i^p$ instead of $m = \langle p, i \rangle$, and use two functions $p = ProcId(m_i^p)$ and $i = messageNumber(m_i^p)$ to reveal the process identity and the message number of a message. Using an established notational device we also write $m.\text{id}$ for $ProcId(m)$, and $m.\text{number}$ for $messageNumber(m)$.
>
> We tacitly assume that our language can form pairs, and so for any p and i the pair $\langle p, i \rangle$ exists.
>
> **Events:** These may be terminating or non-terminating.

Predicates: $M\_SendEvent$ and $M\_RecEvent$ are unary predicates defined on the events. The letter M is used to emphasize that these predicates are part of the Message Domain M. If $M\_SendEvent(e)$ holds then we say that e is a send event, and if $M\_RecEvent(e)$ holds then we say that e is a receive event.

Functions: The first four functions in the following list have a-temporal values and the remaining three have event values.

(1) $ProcId : Events \cup Messages \to Processes$.

(2) $messageNumber : Messages \to MessageNumbers$.

(3) $Content : Events \cup Messages \to Data$ (partially defined).

(4) *message* : *Events* → *Messages* (partially defined on terminating *Send/Receive* events only).

(5) *SendMsg* : *Messages* → *Events*.

(6) *RecMsg* : *Messages* × *Processes* → *Events* (partial).

(7) γ : *Events* → *Events*. This is the return function.

The axioms (in this signature) that express the intended meaning of the symbols are displayed in Figure 10.48. (We even have a little illustration for the axioms.) The axioms are all natural and we comment on the second and fifth axioms.

Axiom (2) implies that if if $M\_SendEvent(e_1)$ and $M\_SendEvent(e_2)$ and $m = message(e_1) = message(e_2)$ then $e_1 = e_2$ (= $SendMsg(m)$). This may seem to exclude the possibility of repeated sending of the same message (in case of loss). This is not the case since we view here send events as higher-level events that comprise every lower-level event that may be associated with the sending of m, including repeated sends. This view implies that send events by some process are not necessarily serially ordered. (Since, for $i < j$, message m_i^p may require repeated sending well after m_j^p is sent. Axiom (6) therefore only reflects our assurance that $SendMsg(m_j^p)$ begins before $SendMsg(m_k^p)$ does if $j < k$.

Axiom (5) leaves two possibilities for modeling a network domain, namely to require that $M\_SendMsg(m) \not\prec M\_RecMsg(m, q)$ for any message m and process q for which $M\_RecMsg(m, q)$ is defined, or else to strengthen this to $M\_SendMsg(m) \prec M\_RecMsg(m, q)$. The stronger statement is used by most writers and is quite natural, but there are circumstances where the first seems more suitable. For example when broadcasting a message consists in fact of its repeated sending through all channels. Then it is possible that some process q receives m before it was even sent to some other process.

3. Causality

In an influential paper [19], L. Lamport defined the (potential) causality relation $\longrightarrow$ on the send/receive events in a network, where $a \longrightarrow b$ means intuitively that it is "possible for event a to causally affect event b". Following later writers, causality is defined here on the messages rather than on the events. Given a system execution that satisfies the Message Domain Axioms, we make the following definition.

DEFINITION 3.1. *If m_1, m_2 are messages, then $m_1 <_M m_2$ is a shorthand for $m_1.id=m_2.id$ and $m_1.number < m_2.number$.*

The causal relation $\xrightarrow{causal}$ is defined on the messages as the least transitive relation that satisfies the following:

(1) $m_1 \xrightarrow{causal} m_2$ whenever $m_1 <_M m_2$.

(2) $m_1 \xrightarrow{causal} m_2$ whenever $RecMsg(m_1, q) \prec SendMsg(m_2)$ and $q = m_2.\text{id}$.

In plain words, $m_1 \xrightarrow{causal} m_2$ if m_1 is received by the sender of m_2 before it was sent.

Message Domain Axioms

(1) Every event falls under one and only one of the predicates $M\_SendEvent$ and $M\_RecEvent$. $M\_SendEvents$ are all terminating.

(2) If e is any event such that $M\_SendEvent(e)$, then for some unique message m, $e = SendMsg(m)$, and in fact $m = message(e)$. That is, the function $SendMsg$ is one-to-one from its domain onto the set of $M\_SendEvents$, and $message$ is its inverse.

If $r = RecMsg(m, q)$ is defined, then $M\_RecEvent(r)$ holds, $m = message(r)$, and $q = ProcId(r)$.

If r is any terminating event such that $M\_RecEvent(r)$, then, $m = message(r)$ is defined. That is, any terminating receive event is associated with a (unique) message m. In this case, if $q = ProcId(r)$, then $r' = RecMsg(m, q)$ is defined. (Why $r' = RecMsg(m, q)$ and not $r = RecMsg(m, q)$? Because of the possibility that a message is received more than once. In such a case $RecMsg(m, q)$ would denote the first receive event.)

(3) For every message $m = m_k^p$, if $SendMsg(m) = e$ then $M\_SendEvent(e)$, $p = ProcId(e)$, and $Content(m) = Content(e)$.

(4) For every message m and process q, if $RecMsg(m, q) = r$ is defined then r is terminating, $M\_RecEvent(r)$, $ProcId(r) = q$, and

$$Content(r) = Content(m).$$

(5) We require that $\gamma(r) \not\vdash r$ for every terminating receive event r. (Alternatively, a stronger requirement can be made, that $\gamma(r) \prec r$. A remark is made below on this possibility.)

The return function γ is definable by the relation

$$\gamma(r) = SendMsg(message(r))$$

that holds for every terminating receive event r. Alternatively, one could say that whenever $s = SendMsg(m)$ and $r = RecMsg(m, q)$ is defined, then $s = \gamma(r)$. As a result, γ is not strictly necessary, but we find it handy to have it in our signature.

(6) For each process p the messages in p are sent in order of their message numbers. That is, for every $j < k$,

$$begin(\, SendMsg(m_j^p)\,) \prec begin(\, SendMsg(m_k^p)\,).$$

FIGURE 10.48. The Message Domain Axioms.

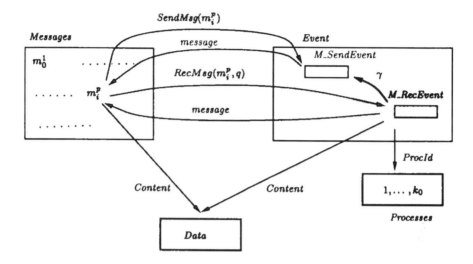

FIGURE 10.49. An illustration for the Message Domain Axioms.

This definition can be given an equivalent explicit formulation. A finite sequence of messages $m_1, m_2, \ldots, m_k$ is said to be a causal sequence iff:

for every index $1 \leq j < k$, if $q = m_{j+1}$.id then $RecMsg(m_i, q) = r$ is defined and $r \prec SendMsg(m_{j+1})$.

Now define $m \xrightarrow{causal} m'$ iff either $m <_M m'$ or there is a causal sequence $\langle m_1, \ldots, m_k \rangle$ such that $m \leq_M m_1$ and $m_k \leq_M m'$.

It is easy to check that these two definitions for $\xrightarrow{causal}$ are equivalent.

The following lemma proves that $\xrightarrow{causal}$ is an ordering relation on the messages, i.e. a transitive and irreflexive relation.

LEMMA 3.2. *If a and b are messages such that $a \xrightarrow{causal} b$, then*

$$begin(\ SendMsg(a)\ ) \prec begin(\ SendMsg(b)\ ).$$

Hence if a and b are by the same sender then $a \xrightarrow{causal} b$ implies $a <_M b$. In particular, $\xrightarrow{causal}$ is cycle free. Another consequence is that for any message m there are only finitely many messages a such that $a \xrightarrow{causal} m$.

Proof. Recall that for higher-level events X and Y, $begin(X) \prec begin(Y)$ means that $x \prec y$ for some lower-level events x and y in X and Y. If I_X and I_Y are the temporal intervals representing X and Y then $begin(X) \prec begin(Y)$ implies $Left\_End(I_X) < Left\_End(I_Y)$. We assume that if for some event e $X \models e \prec Y$, then $begin(X) \prec begin(Y)$.

The lemma thus says that if $a \xrightarrow{causal} b$, then the sending of a must begin before the sending of b. For the proof suppose that $a \xrightarrow{causal} b$ and consider the explicit equivalent definition. If $a <_M b$, then a.id $= b$.id and a.number $< b$.number.

If $e_1 = SendMsg(a)$ and $e_2 = SendMsg(b) = e_2$ then $begin(e_1) \prec begin(e_2)$ (by Axiom 6). Now suppose that there is a causal sequence $\langle m_1, \dots, m_k \rangle$ such that $a \leq_M m_1$ and $m_k \leq_M b$. The proof of the lemma in this case is by induction on k. The simplest case is $k = 2$. Let in this case $u = ProcId(m_2)$ and then

$$e = RecMsg(m_1, u) \prec SendMsg(m_2)$$

by definition. But $SendMsg(m_1) = \gamma(e) \vdash e$ by the fifth Message Domain axiom, and thus the lemma follows in case $a = m_1$, $b = m_2$. If $a \leq_M m_1$ and $m_2 \leq_M b$, then the lemma follows as well since $begin(\ SendMsg(a)\ ) \preceq begin(\ SendMsg(m_1)\ )$ and $begin(\ SendMsg(m_2)\ ) \preceq begin(\ SendMsg(b)\ )$.

Now in case $k > 2$ apply the inductive assumption to the shorter causal sequence $\langle m_2, \dots, m_k \rangle$ and obtain that $begin(\ SendMsg(m_2)\ ) \prec begin(\ SendMsg(b)\ )$. But by case $k = 2$ applied to $\langle m_1, m_2 \rangle$

$$begin(\ SendMsg(a)\ ) \prec begin(\ SendMsg(m_2)\ ).$$

So the lemma follows.

The conclusion that $\xrightarrow{causal}$ is cycle free is obvious now because, if $a \xrightarrow{causal} a$, then $begin(\ SendMsg(a)\ ) \prec begin(\ SendMsg(a)\ )$ which is impossible.

The final remark is a consequence of the finiteness property. For any terminating event, and in particular for $s = SendMsg(m)$, there are only finitely many events x such that $x \vdash s$. $a \xrightarrow{causal} m$ implies $SendMsg(a) \vdash SendMsg(m)$ and hence there are only finitely many messages a such that $a \xrightarrow{causal} m$.

NOTE 3.3. *If $RecMsg(m, q)$ is defined for every process q, then in fact a stronger form of the finiteness property holds. Namely, there is a finite set F of messages such that $m \xrightarrow{causal} n$ for every message $n \notin F$.*

4. Causality preservation and deliveries

Let S be a message domain. That is, a system execution that satisfies the Message Domain axioms. Roughly speaking, we say that the causality relation is preserved in S if whenever m_1 and m_2 are messages such that $m_1 \xrightarrow{causal} m_2$ and q is a process such that $RecMsg(m_1, q)$ and $RecMsg(m_2, q)$ are defined, then the user at process q obtains m_2 only after m_1.

Several researchers have stressed the importance of causality preservation but some think that causality is not of basic importance.[1] Anyhow, I think that it is an interesting concept which should be investigated.

Causality preservation is achieved by interposing a protocol layer between the network and the users. In addition to *Send* and *Receive* operations, this layer supports a third operation *Deliver* which transfers the messages to the users in their causal ordering. That is, though the *Receive* operation may not respect causality, the delivery protocol buffers the received messages and only passes

---

[1]See D. R Cheriton and D. Skeen, *Understanding the limitations of causally and totally ordered communication*, 14 Sym. on Operating System Principles, Dec. 1993, pp.44–57. K. P. Birman, *A response to Cheriton and Skeen's criticism of causal and totally ordered communication*, ACM Operating System Review 28 (1994), no 1, 11–21.

Delivery Axioms

(1) If $d = DeliverMsg(m, q)$ is defined then:
  (a) $m = message(d)$, $q = ProcId(d)$. (Hence $DeliverMsg$ is one-to-one.)
  (b) $Content(d) = Content(m)$,
  (c) $M\_DeliverEvent(d)$,
  (d) d is terminating, and
  (e) $r = RecMsg(m, q)$ is defined as well, and $r \prec d$. That is, receiving a message is a necessary condition for delivery, and the receipt of a message by a process precedes its delivery.

(2) $DeliverMsg$ is onto the $M\_DeliverEvents$. That is: For any event d such that $M\_DeliverEvent(d)$, if $m = message(d)$ and $q = ProcId(d)$, then $d = DeliverMsg(m, q)$. Hence any message is delivered at most once by a process.

FIGURE 10.50. Delivery Axioms.

them to the users in the correct ordering. From the point of view of the users then, *delivery* is the receiving operation, and *Receive* is a servicing procedure to which they have no immediate access. We are going to be more formal and define an extension of the Message Domain signature in which delivery appears as well.

DEFINITION 4.1. *Message Domain with Delivery is the signature obtained by adding to the Message Domain signature a new unary predicate M_DeliverEvent defined on the events, and a new partial function $DeliverMsg$: $Messages$ × $Processes \to Events$.*

If $M\_DeliverEvent(d)$ holds, then we say that d is a delivery event. If $d = DeliverMsg(m, q)$ then we say that event d is the delivery of message m to process q. $DeliverMsg(m, q)$ is not necessarily defined for every message m and process q, since not every message is necessarily delivered to all processes.

The Delivery Axioms are obtained by adding the axioms of Figure 10.50 to the list of Message Domain Axioms (Figure 10.48) modified by replacing (1) there to

> Every event falls under one and only one of the predicates $M\_SendEvent$, $M\_RecEvent$, and $M\_DeliverEvent$. The *message* function is defined on the $M\_DeliverEvents$ as well. $M\_SendEvents$ are terminating.

Let S be a Message Domain with Delivery (a system execution that satisfies the Message Domain with Delivery axioms). There are now two possibilities to define the causality relation on the messages: The first is the $\xrightarrow{causal}$ relation as defined before, with respect to the $RecMsg/SendMsg$ events. From the point of view of the user, this relation seems to be irrelevant since the user is not aware of the $RecMsg$ events which happen at the lower servicing layer. Thus we are led to the following definition of $\xrightarrow{D}$ which relates to the $DeliverMsg/SendMsg$ pair (D is for Delivery).

DEFINITION 4.2. *Relation* $\xrightarrow{D}$ *is defined on the messages as the least transitive relation such that:*

(1) *If* $m_1 <_M m_2$ *then* $m_1 \xrightarrow{D} m_2$

(2) *If* $DeliverMsg(m_1, q) \prec SendMsg(m_2)$, *and* $q = m_2.id$, *then* $m_1 \xrightarrow{D} m_2$.

DEFINITION 4.3. *Causality preservation can be understood in two ways:*

(1) *By causality preservation at the protocol layer we mean that* $m_1 \xrightarrow{causal} m_2$ *implies* $DeliverMsg(m_1, q) \prec DeliverMsg(m_2, q)$ *whenever both events are defined.*

(2) *By preservation of causality at the user layer we mean that* $m_1 \xrightarrow{D} m_2$ *implies* $DeliverMsg(m_1, q) \prec DeliverMsg(m_2, q)$ *whenever these events are both defined.*

Clearly, from the point of view of the users, it is the second definition which is relevant, and yet (in Chapter 11) we shall prove that the first holds in our system. This explains the need for our discussion and for the following lemma.

LEMMA 4.4. *Preservation of causality at the protocol layer implies its preservation at the user layer.*

Proof. Assume preservation of causality at the protocol layer. The lemma follows immediately from the implication

$$\text{If } m_1 \xrightarrow{D} m_2, \text{ then } m_1 \xrightarrow{causal} m_2$$

which we prove now. We have to show that $\xrightarrow{causal}$ satisfies the requirements in the definition of $\xrightarrow{D}$, and then, since $\xrightarrow{D}$ is defined as the least transitive relation that satisfies these requirements, $\xrightarrow{D} \subseteq \xrightarrow{causal}$ follows. The main point is to observe that if $DeliverMsg(m, q) \prec SendMsg(m)$ then $m_1 \xrightarrow{D} m_2$. Indeed $RecMsg(m_1, q) \prec SendMsg(m_2)$ follows, as $RecMsg(m_1, q) \prec DeliverMsg(m_1, q)$. $\square$

A stronger form of causality preservation can also be found in the literature. For the protocol layer, for example, this form is:

If $m_1 \xrightarrow{causal} m_2$ and $DeliverMsg(m_2, q)$ is defined, then $DeliverMsg(m_1, q)$ is defined as well and $DeliverMsg(m_1, q) \prec DeliverMsg(m_2, q)$.

I have chosen the simpler definition, but it may be the case that this form is more useful.

4.1. Uniform deliveries. Even if the network satisfies the FIFO property and preserves causality, a disordering of messages is possible when more than two processes are involved. It is better to illustrate this problem with an example.

Suppose four processes a, b, c, d that only use point-to-point connections. To broadcast a message, a process just sends it to the three other processes, one after the other, using the available channels. Now it is not difficult to find a scenario in which two messages, one broadcast by a and the other by b, arrive to c in a certain order but to d in the reverse order. This is undesirable, and we shall deal here with the question of ensuring a uniform ordering of the messages. For example, a may send message m_1 first to c and then to d and to b, but with a long delay in between the sending to c and the completion of the broadcasting. Suppose that during this delay c receives m_1 and b transmits its message m_2 to c, to d, and then to a. Afterwards a resumes its broadcasting, and both b and a receive messages m_1 and m_2. In this scenario c receives first m_1 and then m_2, but d receives these messages in the reverse ordering. Causality is preserved because no message is received by a or b before sending.

DEFINITION 4.5. *The Uniform Delivery Property is specified in the Message Domain with Delivery signature as the following property of message domains. There is a linear ordering $\lhd$ on the messages such that*

(1) *if* $m_1 \xrightarrow{causal} m_2$ *then* $m_1 \lhd m_2$ *(that is $\lhd$ extends the causal ordering), and*

(2) *if* $m_1 \lhd m_2$ *then for any process p for which $d_2 = DeliverMsg(m_2, p)$ is defined, $d_1 = DeliverMsg(m_1, p)$ is also defined and*

$$d_1 \prec d_2.$$

Uniform Delivery does not assume that all messages are necessarily delivered. In situations where all messages are delivered (as in the following chapter) item (2) becomes

If $m_1 \lhd m_2$ then $DeliverMsg(m_1, p) \prec DeliverMsg(m_2, p)$ for every p.

Note that Uniform Delivery implies causality preservation: If $m_1 \xrightarrow{causal} m_2$ then (1) and (2) imply $DeliverMsg(m_1, p) \prec DeliverMsg(m_2, p)$ whenever the delivery events are defined.

5. Time-stamp vectors

Vectors of time stamps are used to obtain causality preservation. We describe in this section an abstract setting in which time-stamp vectors are used, and in the following chapter a protocol realizes this idea.

The Time Stamp Vector Signature is obtained by adding to the Message Domain with Delivery signature the following sort, predicate, and function.

(1) *TimeStampVector* is a new sort, interpreted as the set of all vectors (sequences) of length k_0 of *MessageNumbers* $\cup \{-1\}$ (recall that k_0 is the number of processes). So, if $v \in TimeStampVector$, then for every $1 \leq j \leq k_0$, $v[j] \geq -1$ is the jth entry of v. The special value, -1, stands for the "undefined" message, and its precise role will be clarified later.

(2) A partial ordering $\leq$ is defined on the time-stamp vectors by

$$v \leq w \text{ iff for every } 1 \leq j \leq k \ v[j] \leq w[j].$$

If $v \leq w$ and $v \neq w$, then we write $v < w$.

(3) A function symbol $ts : Messages \rightarrow TimeStampVector$. ts is defined on the messages and it takes values that are time stamp vectors.

Time Stamp Vector Axioms

(1) For every $1 \leq p \leq k_0$ and $i \in MessageNumbers$,

$$\text{if } v = ts(m_i^p) \text{ then } v[p] = i.$$

That is,

$$ts(m)[m.\text{id}] = m.\text{number}.$$

(2) If $i < j$, then $ts(m_i^p) < ts(m_j^p)$.

(3) Suppose $m, n \in Messages$, $q = n.\text{id}$, and $RecMsg(m, q) \prec SendMsg(n)$. Then

$$\forall k \in Id \ (k \neq q \implies ts(m)[k] \leq ts(n)[k]).$$

(In fact, we shall see that $ts(m) < ts(n)$ can be derived.)

(4) For any message n with $q = n.\text{id}$, if $k \neq q$ is such that $ts(n)[k] \geq 0$ (i.e., $\neq -1$), then there exists a message n' such that

$$RecMsg(n', q) \prec SendMsg(n) \text{ and } ts(n')[k] = ts(n)[k].$$

FIGURE 10.51. The Time Stamp Vector Axioms.

The Delivery Axioms of the previous section are extended with the axioms of Figure 10.51, and they form together the Time Stamp Vector Axioms. The first axiom says that the time-stamp of a message from process p carries at the pth entry its message number. The second axiom says that each process attaches greater time-stamps to its later messages. The third and forth axioms imply, as we are going to prove, that the timestamps reflect the causal ordering. A stronger form of the third axiom (which is proved below) says that the time stamp of any message sent by process q is $>$ than all time stamps that were received prior to the sending. The fourth axiom says that for every message n $ts(n)$ is not determined arbitrarily: at each entry $k \neq n.\text{id}$ either $ts(n)[k]$ is the initial value -1, or else $ts(n)[k] \geq 0$ is determined by some message n', received before the sending. In particular, this axiom implies that if $n = m_i^q$ is not in the domain of $SendMsg$ then except for its qth coordinate $ts(n) = -1$.

LEMMA 5.1. *Assume the Time Stamp Vector Axioms. For every two distinct messages m_i^p, m_j^q, if $i \leq ts(m_j^q)[p]$ then $m_i^p \xrightarrow{causal} m_j^q$.*

Proof. Since for every message n the number of messages n' with $n' \xrightarrow{causal} n$ is finite (Lemma 3.2), the proof can be carried by induction on that number for $n = m_j^q$.

Assume first that $p = q$. Then $ts(m_j^q)[p] = j$ by Axiom (1); so that $i \leq j$ implies $i < j$ (as the messages are distinct) and $m_i^p \xrightarrow{causal} m_j^q$ follows.

Assume now that $p \neq q$. Axiom (4) can be applied to $n = m_j^q$, since $0 \leq i \leq ts(m_j^q)[p]$. This yields a message n' such that $RecMsg(n', q) \prec SendMsg(m_j^q)$ and $ts(n')[p] = ts(m_j^q)[p]$. Hence

$$n' \xrightarrow{causal} m_j^q \text{ and } i \leq ts(n')[p].$$

If $n' = m_i^p$ then $m_i^p \xrightarrow{causal} m_j^q$ follows. Otherwise, induction can be applied to $m_i^p \neq n'$ (since $n' \xrightarrow{causal} m_j^q$), and then $m_i^p \xrightarrow{causal} n'$ follows. Thus $m_i^p \xrightarrow{causal} m_j^q$ as required. $\square$

It follows from this lemma that ts is one-to-one. Assume $ts(m) = ts(n)$ but $m \neq n$. Let $p = m.$id. Then $ts(m)[p] = ts(n)[p]$ implies $m \xrightarrow{causal} n$ by our lemma, and similarly $n \xrightarrow{causal} m$ is obtained, which is impossible by our result that $\xrightarrow{causal}$ is irreflexive.

THEOREM 5.2. *Assuming the Time Stamp Vector Axioms, for every two messages $m_i^p \neq m_j^q$*

$$m_i^p \xrightarrow{causal} m_j^q \text{ iff } ts(m_i^p) < ts(m_j^q) \text{ iff } i \leq ts(m_j^q)[p].$$

Proof. Assume $m = m_i^p \xrightarrow{causal} m_j^q = n$. By definition of $\xrightarrow{causal}$ either $m <_M n$ or there is a causal sequence $\langle m_1, \ldots, m_\ell \rangle$ such that $m \leq_M m_1$ and $m_\ell \leq_M n$. The desired conclusion, $ts(m_i^p) < ts(m_j^q)$ follows from:

(26)   If $n_1 <_M n_2$ are two messages in P_p, then $ts(n_1) < ts(n_2)$,

(this is the second axiom, (2) in Figure 10.51), and

(27)   For $i < \ell$, $ts(m_i) < ts(m_{i+1})$.

Proof of (27). Let $r = m_{i+1}.$id. Since m_i, m_{i+1} are successive messages in a causal sequence $RecMsg(m_i, r) \prec Send(m_{i+1})$, and Axiom (3) gives for every $k \neq r$ that $ts(m_i)[k] \leq ts(m_{i+1})[k]$. So it suffices to prove that

$$ts(m_i)[r] < ts(m_{i+1})[r]$$

to conclude that $ts(m_i) < ts(m_{i+1})$. Assume on the contrary that

(28) $ts(m_i)[r] \geq ts(m_{i+1})[r]$.

As $r = m_{i+1}.$id, $ts(m_{i+1})[r] = m_{i+1}.$number, by Axiom (1). So that (28) implies (by the lemma above) that $m_{i+1} \xrightarrow{causal} m_i$. Yet this contradicts $m_i \xrightarrow{causal} m_{i+1}$ since we proved that $\xrightarrow{causal}$ is irreflexive.

Now the proof of the theorem is concluded:

(1) $ts(m_i^p) < ts(m_j^q)$ clearly implies $i \leq ts(m_j^q)[p]$ (by definition of the time-stamps ordering and as $ts(m_i^p)[p] = i$),

(2) $i \leq ts(m_j^q)[p]$ implies $m_i^p \xrightarrow{causal} m_j^q$ by our lemma. $\square$

It follows from this theorem that in the definition of causality preservation (Definition 4.3) we may replace $m_1 \xrightarrow{causal} m_2$ with $ts(m_1) < ts(m_2)$. Thus preservation becomes:

For every messages m_1, m_2, if $ts(m_1) < ts(m_2)$ then for any process $1 \leq q \leq k_0$

$$Deliver\,Msg(m_1, q) \prec Deliver\,Msg(m_2, q)$$

whenever these events are defined.

11
Uniform delivery in group communication

In group communication a sending process can broadcast a message in a single operation, but messages from different senders may reach the receivers in different orderings. The purpose of this chapter is to explain two solutions to this uniform ordering problem: the All-Ack Protocol and the Early Delivery Protocol (based on Dolev, Kramer, and Malki [14]). An index for this chapter is at page 231

Since we shall use some of the notions that were introduced in the previous chapter, the reader is encouraged to reread that chapter, or to refer to the exact definitions whenever the relevant notion is employed (non-lossy network, causal ordering, uniform ordering etc).

This chapter explains and proves the correctness of a protocol that ensures the Uniform Delivery Property (Definition 4.5 in Chapter 10), namely that the messages are delivered to all the receivers in the same order that extends the causal ordering. The protocol presented here has two variants: the simpler is the All-Ack Protocol and the second is an adaptation of the "Early Delivery" protocol [14]. It will take us a while before we describe the protocols, because we must explain first the data structures which are quite complex. So we begin with a generic protocol in which the complete definition of the data structures is not given. Instead, the abstract properties of these data structures is specified (properties which suffice for the protocols' correctness). Then we describe two possible implementations, which yield the All-Ack and Early Delivery protocols.

1. A generic Uniform Delivery Protocol

The generic protocol implements three operations: *Send, Receive* and *Deliver,* in such a way that the higher-level events satisfy the Uniform Delivery and hence the Causality Preservation requirements. It is a generic protocol in the sense that it actually defines a family of protocols, each obtained by a different implementation of the external operations.

There are k_0 processes in the network: $P_1, \ldots, P_{k_0}$. Each P_p activates two

---

procedure P_p
concurrently do
(1) **repeat_forever**
   (a) produce_a_Data(d);
   (b) $Send_p(d)$
(2) **repeat_forever**
   $Receive\&Deliver_p$

FIGURE 11.52. The generic Uniform Delivery procedure P_p,
$1 \leq p \leq k_0$

---

procedure $Send_p(d : Data)$	**procedure** $Receive\&Deliver_p$
var $k : Id$;	**var** x: $N\_Messages$;
$v : TimeStamp\,Vector$;	w: $vector$
$i : MessageNumber$	$k : Id$;
(initially 0)	$b : Data\_Buffer$ (initially empty)
$x : N\_Messages$;	**begin**
begin	(1) $N\_Receive(x)$;
(1) $\langle \forall k \neq p\ v[k] := TSV_p[k]\rangle$;	(2) $w := TSV_p$;
(2) $v[p] := i; i := i+1$;	(3) $\langle \forall k \neq p\ TSV_p[k] := \max\left\{\begin{array}{c} x.vector[k] \\ w[k] \end{array}\right\rangle$
(3) $x := \langle d, p, v\rangle$;	(4) $b := register(b, x)$
(4) $N\_Send(x)$.	(5) **while** $deliverable(b)$ **do**
end	(a) deliver all messages in $deliver(b)$
	(b) $b := mark\_delivered(b)$
	end

FIGURE 11.53. Declarations of $Send_p$ and $Receive\&Deliver_p$.

procedures: $Send_p$ and $Receive\&Deliver_p$. The protocols are declared in figures 11.52 and 11.53. A detailed description follows.

Each P_p produces a data item d and broadcasts it using its copy of the $Send$ procedure. Concurrently, P_p also receives messages from the network (with its $Receive\&Deliver_p$ procedure) and delivers messages to its user in operation $deliver$ (whenever $deliverable(b)$ holds). We assume that producing and delivering are terminating operations. That is, produce_a_Data is a terminating operation and an execution of produce_a_Data(d) in line (1)(a) determines the value of data variable d. Similarly, if $D = deliver(b)$ is a sequence of messages, then the instruction on line 5(a) is terminating. Think of delivering as an operation that loads each message in the sequence $deliver(b)$ onto some buffer, and then signals to the user the presence of that message (I shall say nothing about this signaling here, as my only interest is in proving that the deliver operation satisfies the Uniform Delivery Property).

The given raw network is called N. It supports two operations that are invoked by the $Send/Receive$ procedures: $N\_Receive(x)$ and $N\_Send(x)$, where x is an $N\_Messages$ (a variable and a value respectively). Two predicates, $N\_SendEvent$

and $N\_RecEvent$ are defined on the raw-network events. $N\_SendEvent(e)$ iff event e is an execution of an N_Send instruction (line 4 of $Send_p$), and $N\_RecEvent(e)$ iff e is an execution of an $N\_Receive$ instruction (line 1 of $Receive$ & $Deliver$). It is assumed that these events with their values form a structure for the Send/Receive Network Signature in which N is non-lossy (see Definition 1.1 in Chapter 10). This assumption simplifies somewhat the formulation and proofs of our results. It implies, as we shall see, that every message is received and delivered by all processes.

Our first objective is to define the data structures and registers employed by the protocol.

Data Structures: The Data types are the following.

(1) $Id = \{1, \ldots, k_0\}$. That is, $k \in Id$ if $1 \leq k \leq k_0$ is a natural number. k_0 is the number of processes, and $k \in Id$ is sometimes called a process.

(2) $Data$. These are the messages that are broadcast and delivered.

(3) $MessageNumber$ = Natural numbers.

(4) $Messages = \{m_i^p \mid p \in Id, \ i \in MessageNumber\}$.

(5) $TimeStampVector$ = array $[1..k_0]$ of $MessageNumber \cup \{-1\}$. If v is in $TimeStampVector$ then $v[i]$ for $i \in Id$ denotes the ith entry of v.

(6) $N\_Messages = Data \times Id \times TimeStampVector$. If x is an $N\_Messages$ then we write $x = \langle x.\text{data}, x.\text{id}, x.\text{vector} \rangle$ for its three fields.

If $x \in N\_Messages$, and $p = x.\text{id}$, $\ell = x.\text{vector}[p]$, then $x.\text{message}$ denotes the message m_ℓ^p.

(7) $Data\_Buffer$. A $Data\_Buffer$ is a structure which we shall specify in detail later. It suffices at this stage to know that the protocol ($Receive$& $Deliver$) uses four external operations on data buffers:

(a) $register(b, x)$ is defined when b is a data buffer and x is some $N\_Messages$. This operation returns a $Data\_Buffer$. (See line 4.)

(b) $deliverable$ is a predicate defined on $Data\_Buffer$. That is, if b is a $Data\_Buffer$, then $deliverable(b)$ is true of false. (See line 5.)

(c) If $deliverable(b)$, then $deliver(b)$ returns a non-empty, finite sequence of $Messages$. (See line 5(a).)

(d) $mark\_delivered(b)$ returns a $Data\_Buffer$. (See line 5(b).)

These are terminating operations. They are the heart of the protocol, and a special subsection (1.2) is devoted for their description. Here we shall be concerned with their abstract properties alone.

Registers: Each procedure P_p has a serial register TSV_p that is shared by its procedures $Send_p$ and $Receive$&$Deliver_p$. TSV_p is an array of integers ≥ -1, of length $k_0 - 1$:

$$TSV_p[k], \text{ for } (1 \leq k \leq k_0) \wedge k \neq p.$$

The initial value of TSV_p is -1 at each entry. The value of TSV_p fails to be a $TimeStampVector$ because the pth entry is missing. For this reason we do not use regular assignment instructions such as $v := TSV_p$ (or $Read_{TSV_p}(v)$) when v is a $TimeStampVector$, but rather the explicit

$$\langle \forall k \neq p \ v[k] := TSV_p[k] \rangle$$

which shows exactly the indices involved. Similarly

$$\langle \forall k \neq p \; TSV_p[k] := v[k] \rangle$$

is a write instruction. Variable v is a *TimeStampVector*, so that $TSV_p := v$ would make no sense.

Enclosing them in angular brackets indicates that these are atomic instructions: A mutual-exclusion guarantee is given that they never overlap. Thus instruction (1) in *Send$_p$* and (3) in *Receive&Deliver$_p$* are mutually exclusive.

In fact TSV_p is a local register, written by *Receive&Deliver$_p$* and only read by *Send$_p$*

Let us go over the protocol informally to familiarize ourselves with its operations. After producing a *Data* item d, procedure P_p activates *Send$_p$*. First (in lines 1 and 2) the value of the time-stamp vector v is determined. The pth entry $v[p]$ records the number of activations of *Send$_p$* so far. It is determined by the assignment $v[p] := i$ and the updating $i := i + 1$. The other entries of v are copied from TSV_p in a single atomic operation. Then (in line 3) an *N_Messages* x is formed as a triple $\langle d, p, v \rangle$, and is sent through the N-network (in line 4). Concurrently with 1(a)(b) (producing and sending of messages) process P_p repeats *Receive&Deliver$_p$*. *Receive&Deliver$_p$* obtains an *N_Message* x from the net (line 1), and calculates the maximum at each entry q distinct from p of the value at TSV_p and the value found in x. The writing of this maximal value on TSV_p is done atomically. Now (lines 4-5) come several operations on the data buffer b which we cannot yet understand completely. First, operation $b := register(b, x)$ registers the incoming message $m = x$.message in *Data_Buffer* b. The role of b is to keep the messages until the proper delivery moment has arrived. The determination of that moment is, of course, the central feature of the protocol. If, after the registration of x, some messages buffered in b become deliverable, then (see line 5) the procedure delivers those messages that are returned by *deliver*(b). The data buffer itself is then updated to mark the fact that the messages in *deliver*(b) were delivered. If, after this updating, more messages can be delivered from b, then this procedure is repeated until b is depleted and becomes undeliverable.

A process P_p receives messages from *all* processes, including itself, and it delivers to the user its own messages. In real, working systems it makes no sense for a process to use the net to send self-addressed messages, but self-delivery simplifies somewhat the protocols and the formulation of the properties proved.

The correctness of the protocol, namely the Uniform Delivery Property, depends on the special properties of the operations on the data buffer. However, at this stage, there are some important properties that can be derived without any assumptions on these operations, and we shall prove that the Time-Stamp Vector axioms hold for the protocol in the following subsection. Then, in Subsection 1.2, we define *Data_Buffers*, and specify the four operations on them. This means that we do not yet define the *Data_Buffer* operations, but only list their abstract properties which suffice for the correctness of the protocol. Subsection 1.3 defines a possible set of operations, the All-Ack Protocol, which satisfies the requirements of 1.2. In Section 2 we prove that the protocol achieves Uniform Delivery whenever its operations satisfy the abstract requirements of Subsection

1.2. Then, in Section 3 the Early Delivery Protocol is defined—that is, the Early Delivery operations on *Data_Buffers* are defined.

1.1. The Time-Stamp Vector Axioms are satisfied. Let system execution S be an execution of the generic Uniform Delivery Protocol. First we define the higher-level events and functions, and prove that the Message Domain Axioms hold. Then we shall deal with the Time Stamp Vector axioms.

The network that the protocol uses (in invoking N_Send and $N\_Receive$) is called N. As we have said N is assumed to be non-lossy. As defined below, the protocol implements a higher-level network which is called M. We shall define higher-level events, predicates, and functions corresponding to the Message Domain Signature.

An execution of procedure $Send_p$ is called an $M\ Send_p$ event. If S is such an execution, then we define $M\_SendEvent(S)$, and $ProcId(S) = p$. If $x = \langle d, p, v \rangle$ is the $N\_$message that is sent by S (in executing $N\_Send(x)$ in line (4)) and $i = v[p]$, then we define $m_i^p = message(S)$, $d = Content(m_i^p)$, $S = SendMsg(m_i^p)$, and $v = ts(m_i^p)$. We also write $v = ts(S)$.

Recall (page 190) that each $N\_SendEvent$ is terminating in any $Send/Receive$ network. Hence P_p contains infinitely many $M\_SendEvents$ events.

Now let E be an execution of $Receive\&Deliver_p$. The first three events in E that correspond to the execution of lines (1)–(3) in E are collected to form a higher-level $Receive_p$ event. If R is such an event, then we write $M\_RecEvent(R)$, and $p = ProcId(R)$. So the first event in R is an $N\_Receive$ event, the second is a read of TSV_p, and the third is a write on TSV_p. If $x = \langle d, q, v \rangle$ is the value of the $N\_$message obtained in R, and $m = x$.message (i.e., $m = m_{v[q]}^q$) then we define $m = message(R)$, and $d = Content(R)$. We also define $R = RecMsg(m, p)$ unless process p has already received m (and then $RecMsg(m, p) \prec R$).

Recall the assumption that the underlying net N satisfies the $Send/Receive$ axioms, and is non-lossy. Thus $\gamma(r) \prec r$ for every $N\_RecEvent\ r$. It follows from this that for any message m, if $S = SendMsg(m)$ and $R = RecMsg(m, p)$ then $S \prec R$. (If r is the $N\_Receive$ event in R, then r is the first event in R, and $\gamma(r)$ is the last event in S.) This is Axiom 5 in the Message Domain Axioms.

We leave it to the reader to check that all the Message-Domain axioms are satisfied, and we continue with the Delivery Axioms.

Return to E (an execution of the $Receive\&Deliver_p$ procedure) and assume that condition $deliverable(b)$ holds when the **while** loop of line (5) is approached. Then the body of the loop is executed at least once, and each call to $deliver(b)$ returns a finite sequence of messages which are delivered to the user. Each delivery of a message m in this sequence forms what we call a $Deliver_p$ event. If D is this event, then we write $M\_DeliverEvent(D)$, and $m = message(D)$. $ProcId(D) = p$, and $Content(D)$ are defined naturally, and the Delivery Axioms are easily checked, if we accept that operations $deliver(b)$ and $mark\_delivered(b)$ ensure in tandem that no message is delivered twice by P_p. (This will be obvious after the next subsection.)

There is no problem in proving that every execution of $N\_Receive$ (in procedure $Receive\&Deliver$) is terminating. This is by now a familiar argument, so we can be a little terse here. Recall that we assume that N is a non-lossy net, and so for every attentive process P_q, every $N\_SendEvent$ s is of the form $s = \gamma(r)$, where r is a terminating $N\_RecEvent$ in P_q. P_q is attentive (executing forever $Receive\&Deliver_q$), and hence γ—restricted to the P_q events—is onto the $N\_SendEvents$. But there are infinitely many $N\_SendEvents$, and hence there are infinitely many $N\_RecEvents$, and each is therefore terminating. Since all the $Data\_Buffer$ operations are terminating, there must be an infinite number of $Receive\&Deliver_p$ events for every p.

NOTE 1.1. *Thus we have proved not only that the Message Domain Axioms are satisfied, but also that $RecMsg(m, P_q)$ is defined for every message m and process P_q.*

The Time-Stamp Vector axioms of Figure 10.51 in Chapter 10 will be verified next.

LEMMA 1.2. *If $e_1 \prec e_2$ are two write events on register TSV_p (executions of line (3) of $Receive\&Deliver_p$) then $Value(e_1) \leq Value(e_2)$. (That is at each entry $k \neq p$, $Value(e_1)[k] \leq Value(e_2)[k]$.)*

Proof. Let $E_0 \prec E_1 \prec \cdots$ be the infinite sequence of $Receive\&Deliver$ executions. Let e_i be the write event on TSV_p in E_i. We shall prove that $Value(e_i) \leq Value(e_{i+1})$ to conclude the lemma. In executing line (2), E_{i+1} reads TSV_p, and, since this is a local register written only by $Receive\&Deliver_p$ operations, it obtains the value written by E_i. Thus the value of variable w in E_{i+1} after the assignment is $Value(e_i)$. Now in executing line (3), E_{i+1} compares this value with the values of x.vector$[k]$ and picks the maximum to be written on TSV_p. Thus $Value(e_1) \leq Value(e_{i+1})$. $\square$

It follows from this lemma, and the seriality of register TSV_p, that if $r_1 \prec r_2$ are two reads of TSV_p, then $Value(r_1) \leq Value(r_2)$.

Let $S_0 \prec S_1 \prec \cdots$ be the infinite serial sequence of the $Send_p$ events. It follows from our definitions that $message(S_i) = m_i^p$, and $ts(m_i^p) = ts(S_i) = v$ is the $TimeStampVector$ v where $x = \langle d, p, v \rangle$ is the value of the N_Send event in S_i.

LEMMA 1.3. *For any $p \in Id$, if $i < j$ then $ts(m_i^p) < ts(m_j^p)$.*

Proof. It follows that $ts(m_i^p)[p] = i$ (see the assignment $v[p] := i$ of line (2)). The rest of $ts(m_i^p)$ is determined in line (1) by reading TSV_p. But, if r_i and r_j are the reads of TSV_p in S_i and S_j where $i < j$, then $Value(r_i) \leq Value(r_j)$ (this was concluded after the last lemma). Hence $ts(m_i^p) < ts(m_j^p)$ follows. $\square$

So far we have obtained the first two axioms in the list of the Time-Stamp Vector Axioms. The following is not difficult to prove.

LEMMA 1.4. *If $R = RecMsg(m, q)$ and e is the write onto TSV_q (in executing line (3)) then for every $k \neq q$ $Value(e)[k] \geq ts(m)[k]$.*

To prove the third axiom assume that

$$R = RecMsg(m, q) \prec S = SendMsg(n)$$

where $q = n.\text{id}$. We have to prove that

$$\forall k \neq q \ (ts(m)[k] \leq ts(n)[k]).$$

Let e be the write onto TSV_q in R (the execution of line (3)) and let r be the read of TSV_q in S (an execution of line (1)). Then $R \prec S$ implies $e \prec r$, and thence $e \preceq \omega(r)$, where $\omega(r)$ is the write onto TSV_q that affects r. By Lemma 1.2, $Value(e) \leq Value(\omega(r)) = Value(r)$. Hence for every $k \neq q$ $Value(e)[k] \leq Value(r)[k]$. By the above lemma (1.4)

$$\forall k \neq q \ ts(m)[k] \leq Value(e)[k].$$

Hence $ts(m)[k] \leq Value(r)[k]$ for $k \neq q$. But $Value(r)$ determines all entries of $ts(n)$ that are different from q, and thence $ts(m)[k] \leq ts(n)[k]$, which proves the third axiom.

The following lemma is employed in proving the fourth axiom.

LEMMA 1.5. *Let e be the write event on TSV_q in some $Receive_q$ event R, and suppose that $Value(e)[k] \geq 0$ for some $k \neq q$. Then there exists a message m such that $M\_RecEvent(m, q) \preceq R$ and $ts(m)[k] = Value(e)[k]$.*

Proof. Let $R_0 \prec R_1 \prec \cdots$ be the sequence of $Receive_q$ events. Then $R = R_\ell$ for some ℓ, and the lemma can be proved by induction on ℓ. Consider e; it is an execution of line (3) in R, and hence

$$Value(e)[k] = \max\{x.\text{vector}[k], w[k]\}.$$

Case 1: $Value(e)[k] = w[k] \geq 0$.
The value of w is determined by the assignment $w := TSV_q$. Since $w[k] \geq 0$, it is not the initial vector of -1 that is obtained in w, and hence this value of TSV_q was written by $R_{\ell-1}$, and induction can be applied to conclude the lemma.
Case 2: $Value(e)[k] = x.\text{vector}[k]$.
Then $m = x.\text{message}$ and $R = M\_RecEvent(m, q)$ prove the lemma. $\square$

To prove the fourth axiom assume that $n = m_j^q$ is a message with $v = ts(n)$. Suppose $k \neq q$ is such that $v[k] \geq 0$. Following the definition of $ts(n)$, let S_j be the jth $Send_q$ event. $n = message(S_j)$ and $v = ts(S_j)$. As $k \neq q$, $v[k]$ is determined in line 1 in S_j in executing $v[k] := TSV_q[k]$. Let r be this read of register TSV_q. As $v[k] \geq 0$, $\omega(r)$ is not the initial write on TSV_q, and thus $\omega(r) \in R$ for some $Receive_q$ event R. Moreover, $\omega(r) = e$ is the last event in R and hence $R \prec S_j$. The conclusion of Lemma 1.5 proves Axiom 4.

1.2. Data Buffers. Data buffers were used unspecified in the generic protocol, and now we define them. A *Data_Buffer* is a three-sorted structure B of the following form that satisfies the requirements enumerated below.

$$B = (Messages, Data, TimeStamp\,Vector,\ Registered, D, Content, ts)$$

where

(1) The universe of B consists of *Messages*, *Data*, and *TimeStamp Vector*, which are the types defined above (page 205).

(2) *Registered* is a unary predicate on *Messages*. That is *Registered* $\subseteq$ *Messages*. When *Registered*(m) holds, we say that m is registered in B.

Data Buffer Axioms

---

(1) Let $\longrightarrow$ be the binary relation on *Messages* defined by

$$m_i^p \longrightarrow m_j^q \text{ iff } m_i^p \neq m_j^q, \; Registered(m_j^q), \text{ and } i \leq ts(m_j^q)[p].$$

Then $\longrightarrow$ is a partial, irreflexive ordering such that $m_i^p \longrightarrow m_j^p$ when m_j^p is registered and $i < j$.

(2) $D \subseteq Registered$, and D is an initial segment of $\longrightarrow$.

(3) The finiteness property holds. For any message m there is a finite set F of messages such that if n is a registered message not in F then $m \longrightarrow n$.

FIGURE 11.54. *Data_Buffer* Axioms.

(3) D is another unary predicate on *Messages*. The messages in D are said to be delivered in B.

(4) $Content : Registered \rightarrow Data$, and $ts : Registered \rightarrow TimeStamp\,Vector$.

The properties of Figure 11.54 are required of B to be a *Data_Buffer*.

The relation $\longrightarrow$ defined in (1) is called "the precedence relation of B". For an arbitrary ts function there is no reason for this definable relation to be an ordering, and that is exactly what the first axiom requires.

The second axiom requires that D is an initial segment of $(Messages, \longrightarrow)$. In general, if $<_P$ is an irreflexive, partial ordering on P (i.e., a transitive relation such that $a <_P b$ implies $a \neq b$ and $a, b \in P$) and if $A \subseteq P$, then A is an initial segment of $<_P$ iff $x <_P y \wedge y \in A$ implies $x \in A$. The union of two or more initial segments of $<_P$ is again an initial segment. We shall use this simple fact in the following situation. Suppose that $A_0 \subseteq P$ is an initial segment and $A_0 \subseteq B \subseteq P$ is any set. Then there is a maximal initial segment A_1 with $A_0 \subseteq A_1 \subseteq B$. This A_1 is defined as the union of all initial segments X that satisfy $A_0 \subseteq X \subseteq B$.

The third axiom states the finiteness condition of $\longrightarrow$; it is a second-order statement of course. It is trivially true if the set of registered messages is finite, since in this case F is just this set. For the infinite case, recall Note 1.1 above, and Note 3.3 on page 196.

All *Data_Buffers* have the same universe, namely the set of all messages, data values, and time stamp vectors. Hence any *Data_Buffer* is characterized by its predicates and functions alone. If B is a *Data_Buffer*, then we write $Registered^B$, D^B, $Content^B$, and ts^B for their interpretations in B. We also write $\longrightarrow_B$ for the relation defined in Figure 11.54.

If $Registered^B$ is empty, then we say that B is an *empty Data_Buffer*. In this case D^B, $Content^B$, and ts^B are also empty.

If $Registered^B = Messages$, then we say that B is *fully registered*.

Examples of fully registered data buffers can be obtained from models of the Time-Stamp Vector Axioms (Figure 10.51, Chapter 10) as follows.

DEFINITION 1.6. *Let S be a system execution that satisfies the Time-Stamp Vector axioms. Define a fully registered Data_Buffer $B = B_S$ by:*

(1) $Registered^B = Messages$ *is the set of all messages.*

(2) $D^B = \emptyset$. *So no message is delivered.*

(3) $Content^B = Content^S$.

(4) $ts^B = ts^S$.

We must check that the *Data_Buffer* axioms hold for B. Axiom (1) follows from Theorem 5.2 (page 201) which states that if $m \longrightarrow n$ is defined by

$$m \neq n \ \wedge \ m.\text{number} \leq ts(n)[m.\text{id}],$$

then $\longrightarrow$ is $\overset{causal}{\longrightarrow}$. Hence $\longrightarrow$ is transitive, irreflexive, and extends $<_M$. (For $<_M$ see Definition 3.1 in page 193.) The finiteness property for $\overset{causal}{\longrightarrow}$ was also proved as a note on page 196 which follows from our Note 1.1. Thence the *Data_Buffer* axioms.

We now turn towards the four operations on data buffers: *register*, *deliverable*, *deliver*, and *mark_delivered*.

Operation *register*(B, x) can be defined now when B is a *Data_Buffer* and x an *N_Messages*, with $x = \langle d, p, v \rangle$. Let $m = m^p_{v[p]} = x.\text{message}$. Let B' be the expansion of B obtained by defining

$$Registered^{B'} = Registered^B \cup \{m\};$$

$$D^{B'} = D^B;$$

$$Content^{B'}(m) = d$$

(and $Content^{B'}(n) = Content^B(n)$ for any other registered message);

$$ts^{B'}(m) = v$$

(and $ts^{B'}(n) = ts^B(n)$ for any other registered message).
If $m \notin Registered^B$ and the resulting structure B' is again a *Data_Buffer*, then we define $B' = register(B, x)$. Otherwise $register(B, x) = B$.

Before continuing with the other operations we shall define a partial ordering on *Data_Buffers*, denoted $\lhd$.

DEFINITION 1.7. *Let B_1 and B_2 be Data_Buffers. Define $B_1 \lhd B_2$ if and only if:*

(1) $Registered^{B_1} \subseteq Registered^{B_2}$.
(2) $D^{B_1} = D^{B_2}$.
(3) $Content^{B_1} \subseteq Content^{B_2}$.
(4) $ts^{B_1} \subseteq ts^{B_2}$.

In simple words, $B_1 \lhd B_2$ iff B_2 expands B_1, but $D^{B_1} = D^{B_2}$.

We can continue now with the remaining three *Data_Buffer* operations and their properties.

The demands on *deliverable* and *deliver* are displayed in Figure 11.55. These are quite natural demands: (1) says that when all messages are registered (and only a finite number has been delivered) then the data buffer is deliverable. (2) says that in fact any delivery is based on a finite information only. That is, whenever a data buffer is deliverable then there is a finite reason for that, and a finite set of registered messages is a witness for that reason. (3) says that deliveries leave no gap and that the delivery ordering respects $\longrightarrow$. (4) speaks about the absoluteness of the delivery sequence.

(1) If B is fully registered and D^B is finite, then *deliverable(B)*.
(2) If *deliverable(B)* then there is a finite set $E \subseteq Registered^B$ such that if C is any *Data_Buffer* such that $C \lhd B$ and $E \subseteq Registered^C$, then *deliverable(C)*.
(3) If *deliverable(B)*, then *deliver(B)* is defined. *deliver(B)* is a finite, nonempty sequence $m_1, \ldots, m_d$ of messages that are registered in B but are not in D^B, such that if $X = \{m_1, \ldots, m_d\}$ is the set of messages then
    (a) $D^B \cup X$ is again an initial set of messages.
    (b) if $a, b \in X$ and $a \longrightarrow b$, then a is enumerated before b (i.e., $a = m_i$, $b = m_j$ with $i < j$).
(4) If $B_1 \lhd B_2$ where B_1, B_2 are deliverable data buffers, then
$$deliver(B_1) = deliver(B_2).$$

FIGURE 11.55. The four requirements on *deliverable* and *deliver*.

The function *mark_delivered* is finally defined: If *deliverable(B)* and

$$deliver(B) = \langle m_1, \ldots, m_d \rangle,$$

then $B' = mark\_delivered(B)$ is obtained from B by expanding D^B to

$$D^{B'} = D^B \cup \{m_1, \ldots, m_d\}.$$

Since *mark_delivered* changes only predicate D it follows that B' is again a *Data_Buffer* (as $D^{B'}$ is an initial segment by requirement (3) of Figure 11.55).

The following lemma is easily proved since $B_1 \lhd B_2$ implies $deliver(B_1) = deliver(B_2)$ when these data buffers are deliverable.

LEMMA 1.8. *If $B_1 \lhd B_2$ are deliverable, then*

$$mark\_delivered(B_1) \lhd mark\_delivered(B_2).$$

We pause with our general development and study the following particular example.

1.3. The All-Ack Protocol. The simplest concrete example for operations that satisfy these abstract requirements of the *Data_Buffer* operations is given by the All-Ack Protocol. To describe it we need the definition of pending and candidate messages in a data buffer B.

Let $m = m_i^p$ be a registered message in B. If m is not delivered but for every $0 \le \ell < i$, m_ℓ^p is delivered, then m is said to be *pending* in B. Thus, in order for m to be pending two conditions must be satisfied:

(1) *Registered(m)*.
(2) $\neg D(m)$, but $D(n)$ for every $n <_M m$.

If m is pending in B and, for any message n such that $n \longrightarrow m$, n is delivered in B, then m is said to be a *candidate* message in B. A candidate is thus pending, but not every pending message is necessarily a candidate. For m to be a candidate two conditions must be satisfied:

(1) *Registered(m)*, and

(2) $\neg D(m)$, but $D(n)$ for every n such that $n \longrightarrow m$.

Now we can define now *deliverable* and *deliver* for the All-Ack Protocol.

(1) A *Data_Buffer* B is deliverable
iff each process contains a pending message in B.

(2) *deliver*(B) in this case is defined as follows. Let $m^1, \ldots, m^{k_0}$ with m^k in P_k be the set of pending messages in B. Let $X \subseteq \{m^1, \ldots, m^{k_0}\}$ be the maximal subset such that $D^B \cup X$ is an initial segment of $\longrightarrow$. X is clearly non-empty since all candidates are there. Assume some topological-sort algorithm, and let $<_X$ be the resulting linear ordering on X (which respects $\longrightarrow$). For example, for every $m \in X$ let $p(m) = \#\{k \in X \mid k \longrightarrow m\}$ be the number of messages in X that precede m, and define $m <_X n$ iff $p(m) < p(n)$ or else $p(m) = p(n)$ and $m.\mathrm{id} < n.\mathrm{id}$. Now *deliver*$(B)$ is the sequence obtained by enumerating X in the order $<_X$.

We want to prove that the requirements of Figure 11.55 hold for these All-Ack operations.

(1) If B is fully registered but D^B is finite, then for every process p there is a least $i = i(p)$ such that $m_i^p \notin D^B$. Then m_i^p is pending for every $p \in Id$, and hence *deliverable*(B).

(2) Assume *deliverable*(B), and let $E = \{m^1, \ldots, m^{k_0}\}$ with $m^k \in P_k$ be the set of pending messages in B. Clearly, if $C \lhd B$ and $E \subseteq Registered^C$, then E is the set of pending messages in C as well, since $D^C = D^B$. Hence C is deliverable.

(3) If *deliverable*(B) then *deliver*(B) is non-empty because there are candidates and all candidates are in *deliver*(B). Conditions (a) and (b) follow directly from the definition of *deliver*(B).

(4) If $B_1 \lhd B_2$ and B_1 is deliverable, then $D^{B_1} = D^{B_2}$ implies that B_2 is deliverable and

$$deliver(B_1) = deliver(B_2).$$

2. Correctness of the generic protocol

Suppose that the *Data_Buffer* operations satisfy the four requirements of Figure 11.55, and we shall prove here that the resulting protocol satisfies the Uniform Delivery Property (Chapter 10, Definition 4.5). We first define a uniform ordering of messages on any fully registered *Data_Buffer*, and then prove that the deliveries in our protocol are in accordance with this ordering. Let $B = B_0$ be a fully registered *Data_Buffer* such that its set of deliverable messages $D_0 = D^{B_0}$ is empty. Inductively, define a sequence, B_i, called the derived sequence of B_0 by

(29) $$B_{i+1} = mark\_delivered(B_i)$$

The point is that B_i is fully registered and with a finite set of delivered messages, $D_i = D^{B_i}$, and thence B_i is deliverable and B_{i+1} is defined. B_{i+1} is again a fully registered *Data_Buffer* with a finite set of delivered messages, and such that

$$D_{i+1} = D_i \cup deliver(B_i).$$

(One must understand $deliver(B_i)$ in this equation as a set rather than a sequence, of course.)

LEMMA 2.1. Let $D_\infty = \bigcup_{i \in \mathbb{N}} D_i$. Then $D_\infty = Messages$. That is, every message is delivered in some D_i.

Proof. Since $deliver(B_i) = D_{i+1} \setminus D_i$ is non-empty, D_∞ is infinite. So D_∞ is an infinite initial segment of $\longrightarrow$. Given any message m, there is a message $n \in D_\infty$ (by the finiteness property) such that $m \longrightarrow n$. Hence $m \in D_\infty$ as well. $\square$

This decomposition induces a linear ordering $<_B$ of all messages defined as follows. For any message m let $i = i(m)$ be the least integer i such that $m \in D_i$. Since $D_0 = \emptyset$, $i(m) > 0$, and so $m \in D_i \setminus D_{i-1}$ and m is in $deliver(B_{i-1})$. We now define $m_1 <_B m_2$ iff $i(m_1) < i(m_2)$ or else $i(m_1) = i(m_2) = i_0 + 1$ and m_1 is enumerated before m_2 in $deliver(B_{i_0})$.

LEMMA 2.2. $<_B$ extends $\longrightarrow$.

Proof. Suppose that $m \longrightarrow n$ and we shall prove that $m <_B n$. Observe first that $m \longrightarrow n$ implies $i(m) \leq i(n)$: Since $D_{i(n)}$ is an initial segment and $n \in D_{i(n)}$ $m \longrightarrow n$ implies $m \in D_{i(n)}$. Hence $i(m) \leq i(n)$ by the minimality of $i(m)$.

So, in case $i(m) < i(n)$, $m <_B n$ follows. But if $i(m) = i(n)$ then $m <_B n$ follows again since $m \longrightarrow n$, and as the enumeration of *mark_delivered* respects $\longrightarrow$. $\square$

Let S be an execution of the generic Uniform Delivery Protocol. We have proved that the Time Stamp axioms hold in S, and thence $B = B_S$ (defined in Definition 1.6) is a *Data_Buffer*—called the global data buffer of S. Let B_i, for $i \in \mathbb{N}$, be the derived sequence of B. Our aim is to prove that the Uniform Delivery Property holds in S with $<_B$ as the uniform message ordering. In fact, we shall prove that for any $p \in Id$ all messages are delivered by P_p, and

If $m_1 <_B m_2$ then $DeliverMsg(m_1, p) \prec DeliverMsg(m_2, p)$.

In very general terms, the proof consists in observing that each process develops its own copy of B_S and delivers the messages in accordance with the $<_{B_S}$ ordering.

Let P_p be one of the processes (executing the procedure in Figure 11.52) and consider the values taken by *Data_Buffer* variable b in procedure $Receive\&Deliver_p$. The initial value of b is the empty *Data_Buffer*, in which no message is registered. The value of b is modified only by operations *register* and *deliver* (in lines 4 and 5). Let $\langle b_i \mid i \in \mathbb{N} \rangle$ be the sequence of values assumed by b, in the ordering that they are obtained. We wish to compare these values with the derived sequence of B, $\langle B_i \mid i \in \mathbb{N} \rangle$.

LEMMA 2.3. For every i there is a j such that $b_i \lhd B_j$.

Proof. The proof is by induction on i. For $i = 0$, $b_0 \lhd B_0$ because b_0 is the empty *Data_Buffer* and D^{B_0} is empty by definition of $B_0 = B_S$. Now suppose that $b_i \lhd B_j$. This means, by definition, that B_j expands b_i and $D^{b_i} = D^{B_j}$. Look at b_{i+1}. There are two cases.

Case 1: $b_{i+1} = register(b_i, x)$ is a registration, that is, obtained by executing line 4 in some $Receive\&Deliver_p$ event R. Let $m = x.message$ be the message received by R. If m is already in $Registered^{b_i}$, then $b_{i+1} \doteq b_i$, and the inductive claim is obvious in this case. Otherwise, b_{i+1} is obtained by "registering" m without changing D^{b_i}, and $b_{i+1} \lhd B_j$ follows, as B_j is fully registered.

Case 2: $b_{i+1} = mark\_delivered(b_i)$ is obtained by an execution of line 5 in some $Receive\&Deliver_p$ execution for which condition $deliverable(b_i)$ holds. Since $b_i \lhd B_j$ by induction, $b_{i+1} \lhd B_{j+1} = mark\_delivered(B_j)$ by Lemma 1.8. $\Box$

LEMMA 2.4. *There are infinitely many indices i for which*

$$b_{i+1} = mark\_delivered(b_i).$$

If $\langle i(n) \mid n \in \mathbb{N} \rangle$ is an increasing enumeration of this infinite set, then $b_{i(n)} \lhd B_n$ can be proved by induction. Hence $deliver(b_{i(n)}) = deliver(B_n)$ (by property 4).

Proof. Given any $i_0 \in \mathbb{N}$ we want to find $i \geq i_0$ such that

$$b_{i+1} = mark\_delivered(b_i).$$

We know that for some j_0 $b_{i_0} \lhd B_{j_0}$. We also know that B_{j_0} is deliverable (being fully registered with a finite D) and that $B_{j_0+1} = mark\_delivered(B_{j_0})$. So, by requirement (2) of Figure 11.55, there is a finite set $E \subseteq Registered^{B_{j_0}}$ such that for any $Data\_Buffer$ C, if $E \subseteq Registered^C$ and $C \lhd B_{j_0}$, then C is deliverable. Since every message is received (and registered), for some $i \geq i_0$ b_i has registered all messages of E. Assuming that for no j with $i_0 \leq j < i$ $b_{j+1} = mark\_delivered(b_j)$ then only registrations may have changed b_{i_0} so that $D^{b_{i_0}} = D^{b_i}$, and hence $b_i \lhd B_{j_0}$. Thence b_i is deliverable and so the lemma follows. $\Box$

It follows from this lemma that the sequence of messages delivered by P_p is the sequence of all messages in the $<_B$ ordering. Indeed, if we write $d_n = deliver(b_{i(n)})$, then $d_n = deliver(B_n)$. The $<_B$ ordering of the messages puts the messages of d_n before those of d_m if $n < m$, and orders the messages of each $d_n = deliver(B_n)$ in accordance with $\longrightarrow$. But this is exactly the delivery ordering in P_p. If $n < m$ then the messages of $d_n = deliver(b_{i(n)})$ are delivered before those of $deliver(b_{i(m)})$ (as $i(n) < i(m)$), and the messages in $deliver(b_{i(n)})$ are delivered in the sequence ordering of $deliver(B_n)$.

3. The Early Delivery Protocol

To deliver even a single message, the All-Ack Protocol must wait for k_0 undelivered messages to be registered. This may slowdown the system if one of the processes is much slower than the others. A remedy to this problem (as Dolev, Kramer, and Malki [14] have shown theoretically and experimentally) is with "early delivery" protocols that allow deliveries in some situations in which the All-Ack Protocol would have to wait. This section describes the data structure FISH employed in the Early Delivery Protocol and used to define the data-buffer operations.

Let B be a $Data\_Buffer$ with a finite set of delivered messages D^B. We say that message m_ℓ^p is "first undelivered" in B if $m_\ell^p \notin D^B$ but $\forall i < \ell$ $m_i^p \in D^B$.

A first undelivered message m_ℓ^p is registered in B if and only if it is a pending message (but a first undelivered message may well be unregistered). Since D^B is finite there is for every $p \in Id$ a first undelivered message $m^p = m_\ell^p$ in P_p (ℓ depends on p of course). Let $M = \{m^p \mid p \in Id\}$ be the set of all first undelivered messages. Let $E^B \subseteq M$ be the set of all first undelivered messages $m^j \in M$ such that m^j is registered in B and whenever $m^i \longrightarrow m^j$ then m^i is registered as well. The transitivity of $\longrightarrow$ implies that if $n \in E^B$, $m \in M$, and $m \longrightarrow n$, then $m \in E^B$. E^B is called the "evidence set" of B. So the evidence set is an initial segment of M. Clearly every candidate of B is in E^B, because if m is a candidate, then m is registered and minimal in M.

Let $R \subseteq Id$ be defined by $k \in R$ iff $m^k \in E^B$. A structure $F = fish(B)$ (called the fish of B) is defined as

$$F = (Id, R, \longrightarrow)$$

where $\longrightarrow$ is defined on R by $i \longrightarrow j$ iff $m^i \longrightarrow m^j$. Since we want to study these structures in isolation we shall give an independent definition.

DEFINITION 3.1. *A "FISH" is a structure $F = (Id, R, \longrightarrow)$ where: $Id = \{1, \ldots, k_0\}$ is the universe of "processes", $R \subseteq Id$ is a predicate (its members are called "registered processes"), and $\longrightarrow$ is a partial (irreflexive) ordering on R, called "causal ordering". (By saying that $\longrightarrow$ is on R I mean that $a \longrightarrow b$ implies $a, b \in R$.)*

If $p_1 \longrightarrow p_2$ then we say that p_2 "follows" p_1 (or that p_1 precedes p_2).

If $k \in R$, we say that k is registered in F, and otherwise it is unregistered.

Minimal registered processes are called candidates. That is, $k \in Id$ is a candidate process iff $R(k)$ and $\forall k' \in Id \neg(k' \longrightarrow k)$. If R is non-empty, then there must be a candidate. If $k \in R$ is not a candidate, $j \longrightarrow k$ for some j and we say that k *is secondary*.

It is not difficult to check that if B is a *Data_Buffer* and $F = fish(B)$ its fish, then a first-undelivered message m^j is a candidate message (in the sense defined in page 212) iff $j \in R$ is a candidate in F.

FISHES are used by the protocols to determine when and which messages to deliver, and for this the following definitions are required. Let F be a FISH. R^F (also denoted $r(F)$) is the set of registered processes, and $\overline{r}(F) = Id \setminus r(F)$ is the set of non-registered processes in F: the cardinality of this set is $\#\overline{r}(F)$. We say that F is *full* iff $r(F) = Id$, that is $\#\overline{r}(F) = 0$. Thus a FISH is full just in case every process is registered

The set of candidate processes in F is denoted $c(F)$. The number of candidate processes in F is denoted $\#c(F)$ (or just $\#c$ when the identity of F is clearly determined).

The set of non-candidate processes in F is denoted $\overline{c}(F) = Id \setminus c(F)$. Of prime importance is the number

$$\#\overline{c}(F) = k_0 - \#c.$$

$0 \leq \#\overline{c}(F) \leq k_0$ is the number of non-candidate processes. (Think of it as "the uncertainty number of F" to clarify the intuition.)

Fix a real number $0 \leq \alpha < 1$ and define for any FISH F

$$\Phi_\alpha(F) = \#\overline{c}(F) \times \alpha.$$

Φ_α is used below "to determine majority". For definiteness you may set $\alpha = 1/2$ and let $\Phi(F) = n(F) \times 1/2$ in what follows. We write Φ_F instead of $\Phi(F)$ for readability.

For any candidate process m define a subset $V(m)$ of the secondary (registered, non-candidate) processes. $V(m)$ is the set of all secondary processes i that follow m. Formally

$$V(m) = \{i \mid i \in R \text{ and } m \longrightarrow i\}.$$

To emphasize that this definitions is relative to F we may write $V_F(m)$.

If m is a candidate in F and

$$|V(m)| \geq \Phi_F$$

then we say that m "is a source", and if

$$|V(m)| < \Phi_F - \#\bar{r}(F)$$

then we say that m is a "excluded" in F.

Clearly a process cannot be both a source and excluded. It is possible for a candidate to be neither excluded nor a source, but if F is full then any candidate m is either excluded or a source, as $\#\bar{r}(F) = 0$. Since we shall need this result later on, we mark it down as a lemma:

LEMMA 3.2. *If F is full then any candidate process is either excluded or a source (but never both).*

We now make the crucial definitions of deliverable FISHES and deliverable processes. We say that a FISH F is deliverable iff at least one of the following two conditions holds:

(1) $\#\bar{r} \leq \Phi_F$, there exists a source, and each candidate is either a source or else is excluded.

(2) F is full.

It is possible that both conditions hold. If F is deliverable because of the first condition, then we say that it satisfies the *early delivery* condition, and if it satisfies the second condition then we say that it satisfies the *all-ack* condition.

If F is deliverable, then the *deliverable processes* of F are defined in accordance with the following two cases:

(1) If there exists a source in F then the deliverable processes are all sources of F.

(2) If there are no sources in F then the deliverable processes are all candidates of F.

In any case the set of deliverable processes in F is non-empty when F is deliverable.

DEFINITION 3.3 (END-EXTENSION). *If F_1 and F_2 are FISHES (with $F_i = (Id, R_i, \longrightarrow_i)$ for $i = 1, 2$), then F_2 is an end-extension of F_1 iff $R_1 \subseteq R_2$, $\longrightarrow_1 = \longrightarrow_2 | R_1$, and R_1 is an initial segment of $(R_2, \longrightarrow_2)$.*

This ends our somewhat long sequence of definitions and now we draw some consequences.

LEMMA 3.4. *Suppose that F_2 is an end-extension of F_1. Let*

$$newly\_r = r(F_2) \setminus r(F_1)$$

be the set of newly registered processes, and let $\#newly\_r$ be its cardinality. Let

$$new\_c = c(F_2) \setminus c(F_1)$$

be the set of new candidates, and let $\#new\_c$ be its cardinality.

(1) *A registered process in F_1 is a candidate in F_1 iff it is a candidate in F_2. So $\#c(F_1) \leq \#c(F_2)$.*

(2) *For every candidate m in F_1 $V_{F_1}(m) \subseteq V_{F_2}(m)$.*

(3) *Since $\#c(F_1) \leq \#c(F_2)$ and $\#\bar{c} = k_0 - \#c$, $\#\bar{c}(F_2) \leq \#\bar{c}(F_1)$ follows, and hence $\Phi_{F_2} \leq \Phi_{F_1}$. In fact $\Phi_{F_1} - \Phi_{F_2} = \alpha \times \#new\_c$.*

(4) *$V_{F_2}(m) \setminus V_{F_1}(m) \subseteq newly\_r \setminus new\_c$, for every m in $c(F_1)$. If $m \in new\_c$ then $V_{F_2}(m) \subseteq newly\_r \setminus new\_c$.*

(5) *If process m is a source in F_1 then it is a source in F_2, and if it is excluded in F_1 then it is excluded in F_2 as well.*

(6) *Suppose that $\#\bar{r}(F_1) \leq \Phi_{F_1}$, then any new candidate in F_2 is necessarily excluded in F_2.*

(7) *If F_1 is deliverable, then F_2 is deliverable and they have the same set of deliverable processes.*

Proof. The first four items follow directly from the definitions. For example, a candidate m of F_1 remains a candidate because no newly registered process precedes it in any end-extension of F_1.

To prove item (5) suppose first that m is a source in F_1. This means that m is a candidate in F_1 and

$$|V_{F_1}(m)| \geq \Phi_{F_1}.$$

Yet $V_{F_1}(m) \subseteq V_{F_2}(m)$ implies $|V_{F_2}(m)| \geq |V_{F_1}(m)|$. and since it was shown in (3) that $\Phi_{F_1} \geq \Phi_{F_2}$ we get that $|V_{F_2}(m)| \geq \Phi_{F_2}$. Thus m is a source in F_2 too.

We show next that if m is excluded in F_1 then it is excluded in F_2. So assume

$$|V_{F_1}(m)| < \Phi_{F_1} - \#\bar{r}(F_1).$$

We want

$$|V_{F_2}(m)| < \Phi_{F_2} - \#\bar{r}(F_2)$$

which shows that m is excluded in F_2 as well. For that it suffices to prove

$$|V_{F_2}(m) \setminus V_{F_1}(m)| \leq (\Phi_{F_2} - \#\bar{r}(F_2)) - (\Phi_{F_1} - \#\bar{r}(F_1)).$$

Now the right-side of this inequality is $\#\bar{r}(F_1) - \#\bar{r}(F_2) - (\Phi_{F_1} - \Phi_{F_2})$. Clearly

$$\#\bar{r}(F_1) - \#\bar{r}(F_2) = \#newly\_r,$$

and (by 3)

$$\Phi_{F_1} - \Phi_{F_2} = \alpha \times \#new\_c \leq \#new\_c \text{ (as } \alpha \leq 1).$$

Thus it suffices to prove

$$|V_{F_2}(m) \setminus V_{F_1}(m)| \leq \#newly\_r - \#new\_c$$

which is evident from item (4).

To prove the sixth item, assume $\#\bar{r}(F_1) < \Phi_{F_1}$ and consider a new candidate process m in F_2. We claim first that $V_{F_2}(m) \subseteq \bar{r}(F_1)$. If this were not the case then $m \longrightarrow i$ would follow for some i that is registered in F_1. But, as F_2 is an end-extension, this would imply that m is in F_1 as well!

Now we prove that the new candidate m is excluded in F_2. That is

$$|V_{F_2}(m)| < \Phi_{F_2} - \#\bar{r}(F_2).$$

Since (by 4) $V_{F_2}(m) \subseteq newly\_r \setminus new\_c$ (and as $new\_c \subseteq newly\_r$), it suffices to prove that

$$\#newly\_r - \#new\_c < \Phi_{F_2} - \#\bar{r}(F_2).$$

Yet

$$newly\_r = \bar{r}(F_1) \setminus \bar{r}(F_2)$$

and $\bar{r}(F_2) \subseteq \bar{r}(F_1)$) imply that

$$\#newly\_r - \#new\_c = (\#\bar{r}(F_1) - \#\bar{r}(F_2)) - new\_c.$$

As (by 3)

$$\Phi_{F_2} = \Phi_{F_1} - \alpha \times \#new\_c$$

it suffices to prove that

$$(\#\bar{r}(F_1) - \#\bar{r}(F_2)) - \#new\_c < \Phi_{F_1} - \alpha \times \#new\_c - \#\bar{r}(F_2).$$

But this is evident as $\alpha < 1$ and $\#\bar{r}(F_1) \leq \Phi_{F_1}$.

Now we prove the last item of the lemma, and so we assume that F_1 is deliverable. There are two cases: In the first F_1 is full and in the second it is not. Assume first that F_1 is full. Then F_2 is also full and $F_1 = F_2$. Hence F_1 and F_2 have the same deliverable processes.

Now for the second case assume that F_1 is deliverable but not full. The conditions for this early delivery are the following:

(1) $\#\bar{r}(F_1) \leq \Phi_{F_1}$.

(2) There is a source, and every candidate in F_1 is either a source or is excluded.

There are two possibilities to consider:

(1) F_2 is not full. We claim that F_2 satisfies the early delivery criterion as well. The first point to see is that

$$\#\bar{r}(F_2) \leq \Phi_{F_2}.$$

For that it suffices to prove $\#\bar{r}(F_2) - \#\bar{r}(F_1) \leq \Phi_{F_2} - \Phi_{F_1}$ or

(30) $$\#\bar{r}(F_1) - \#\bar{r}(F_2) \geq \Phi_{F_1} - \Phi_{F_2} = \alpha \times \#new\_c.$$

But $new\_c \subseteq newly\_r = \bar{r}(F_1) \setminus \bar{r}(F_2)$ and $0 \leq \alpha < 1$, which imply (30). The second point is that there is a source in F_2, but this is obvious as the source of F_1 remains a source by item 5. The third point is the source-excluded dichotomy for candidates. This follows from item 5 and item 6 which says that every new candidate is excluded in F_2.

(2) F_2 is full. Then by definition F_2 satisfies the all-ack delivery criterion, and is deliverable.

Thus if F_1 is deliverable then F_2 is also deliverable. To see that they both have exactly the same deliverable messages we argue as follows. As F_1 contains a source, a message is deliverable iff it is a source, and then items (5) and (6) in the lemma and the fact that any candidate of F_1 is either a source or excluded imply that F_1 and F_2 have the same sources. $\square$

LEMMA 3.5. *If $B_1 \lhd B_2$ are Data_Buffers, then $F_2 = fish(B_2)$ is an end-extension of $F_1 = fish(B_1)$. If, in addition, $E^{B_2} \subseteq Registered^{B_1}$, then $F_1 = F_2$.*

Proof. Recall Definition 1.7: $B_1 \lhd B_2$ means that

i $Registered^{B_1} \subseteq Registered^{B_2}$,

ii $D^{B_1} = D^{B_2}$,

iii $Content^{B_1}$ is extended by $Content^{B_2}$,

iv ts^{B_1} is extended by ts^{B_2}.

Since $D^{B_1} = D^{B_2}$, B_1 and B_2 have the same set $M = \{m^p \mid p \in Id\}$ of first-undelivered messages. Let $E_i = E^{B_i}$, for $i = 1, 2$, be the evidence sets of B_1 and B_2 (defined on page 216), and let $\longrightarrow_1$, $\longrightarrow_2$ be the precedence relations defined on E_1 and E_2 by item 1 of the *Data_Buffer* Axioms.

CLAIM 3.6.   (1) $E_1 \subseteq E_2$,
(2) $\longrightarrow_1 = \longrightarrow_2 | E_1$.
(3) E_1 is an initial segment of $(E_2, \longrightarrow_2)$.

Proof of claim: Suppose, for (1), that $m^j \in E_1$. Then m^j is registered in B_1 and whenever $m^i \longrightarrow m^j$ in B_1 then m^j is registered in B_1 as well. We shall prove that $m^j \in E_2$. It follows from i above that m^j is registered in B_2 too. Now if $m^i = m^i_\ell$ is any first-undelivered message in B_2 such that $m^i \longrightarrow m^j$ in B_2, then $\ell \le ts^{B_2}(m^j)[i]$. Hence, as $ts^{B_2}(m^j) = ts^{B_1}(m^j)$, $m^i \longrightarrow m^j$ in B_1, which implies that m^i is registered in B_1 and thence in B_2. So indeed $E_1 \subseteq E_2$.

Now we prove (2), that is, for $a, b \in E_1$, $a \longrightarrow_1 b$ iff $a \longrightarrow_2 b$. This follows, again, from the fact that, in data-buffers, $\longrightarrow$ is definable from ts.

Finally, E_1 is an initial segment of E_2; because if $m \longrightarrow_2 n$ where $n \in E_1$ and m is any message in E_2, then $m \longrightarrow_1 n$ as well (since both are equivalent to $m.number \le ts(n)[m.id]$) and hence $m \in E_1$ by the defining property of the evidence set. This proves the claim.

We turn to the lemma. According to Definition 3.3, F_2 is an end-extension of F_1 iff:

(1) $R_1 \subseteq R_2$.
(2) $\longrightarrow_1 = \longrightarrow_2 | R_1$.
(3) R_1 is an initial segment of R_2.

Each item follows from the corresponding item of the claim.

The additional claim follows if we show that $E_2 \subseteq E_1$. But is $E_2 = E^{B_2} \subseteq Registered^{B_1}$, and if $m \in E_2$ then m is registered in B_1. Hence relation $x \longrightarrow m$ has the same definition in E_1 and E_2. Now if $x \longrightarrow m$ in E_1, then x is registered in E_1 by the following argument. First, $x \longrightarrow m$ in E_2. Hence x is in E_2 since the evidence set is an initial segment of M. So x is registered in B_1 by our assumption. $\square$

The Early Delivery predicate and operation *deliverable* and *deliver* will be defined here and proved to satisfy the four requirements of data buffers. If B is a *Data_Buffer*, let $F = fish(B)$ be its FISH. Define *deliverable*(B) iff F is a deliverable FISH, and in that case let $i_1, \ldots, i_d$ be the set of deliverable processes, and define *deliver*(B) as the sequence of messages formed from the set $\{m^{i_1}, \ldots, m^{i_d}\}$ ordered by some topological sorting algorithm that extends $\longrightarrow$. (See the corresponding discussion in subsection 1.3.)

We must check that the four requirements of Figure 11.55 hold for these operations.

(1) If B is a fully registered *Data_Buffer* and D^B is finite, then $R = Id$ (as all messages are registered). Hence $F = fish(B)$ is full and deliverable.

(2) Suppose that B is deliverable, and let $E = E^B$ be its evidence set. By its definition, $E^B \subseteq Registered^B$. Suppose now that $C \lhd B$ is a *Data_Buffer* such that $E^B \subseteq Registered^C$. Then $fish(B) = fish(C)$ (by the last lemma) and hence C is deliverable.

(3) If *deliverable*(B), then $F = fish(B)$ is a deliverable FISH and the nonempty set of deliverable indices of F induces the corresponding set of deliverable messages *deliver*(B). If X is this set of messages, then $D^B \cup X$ is an initial set of messages because the messages in X are all candidates.

(4) If $B_1 \lhd B_2$ are deliverable data-buffers, then $F_1 = fish(B_1)$ is end-extended by $F_2 = fish(B_2)$. Hence F_1 and F_2 have the same deliverable processes (Lemma 3.4(7)) and the same partial ordering, and thus *deliver*$(B_1) = $ *deliver*(B_2) follows.

4. A worked-out example

A situation in which the Early Delivery Protocol seems to have an advantage over the All-Ack Protocol is one in which the normally operating processes send messages which usually reach all other processes after a short delay, but there is also a small number of processes that are momentarily slowed down with the result that their messages take much longer to reach their destinations. The following example illustrates and compares the All-Ack and Early Delivery protocols. We assume eight processes (so $k_0 = 8$); six are operating normally, but P_1 and P_2 are very slow. Figure 11.56 represents 22 messages: Processes P_1 to P_8 contain three messages each, but P_1 and P_2 contain only two messages each. The causal arrows are not shown in order not to encumber the figure. Instead, we make a simplifying (though not very realistic) assumption that the causal precedence is determined by the temporal position of the messages—except for m_0^1 and m_0^2 which are so slow that they precede only the messages in t_6 and t_7. (This is explained in the caption to 11.56.)

First we calculate the All-Ack precedence relation on the messages. We view figure 11.56 as a representation of the global *Data_Buffer* (a partial representation since the global buffer contains an infinite number of messages). The global buffer is fully registered, and with no delivered messages. Let us denote this buffer with B_0^{AA} (where AA is for All-Ack). Messages $m_0^1, m_0^2, m_0^3, m_0^4, m_0^5$ are candidates,

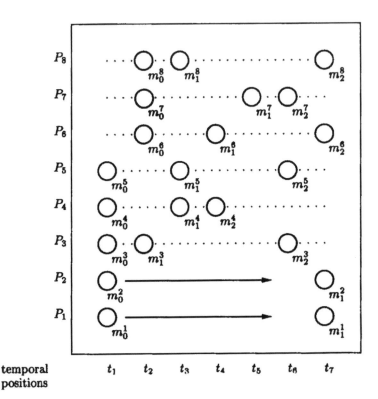

temporal
positions

FIGURE 11.56. Global $Data\_Buffer$ B_0. Messages are represented by circles, where messages on the same line are by the
same process, and each message follows the messages of its process to its left. There are seven "temporal positions", that is,
values on the X-axis. For example, m_2^4 and m_1^6 are in position
t_4. Let $\rightarrow$ be the causal relation on the messages. We assume
here that $\rightarrow$ is simply determined by the temporal position:
$a \rightarrow b$ iff the position of a is strictly before that of b— except
for m_0^1 and m_0^2 which precede only messages on position t_6 and
later. For example $m_0^5 \rightarrow m_0^6$, $m_0^8 \rightarrow m_1^4$, but m_0^7 and m_1^3 are
incomparable, and $m_0^1 \rightarrow m_2^5$ but m_0^2 and m_1^7 are incomparable.
Likewise $m_1^3 \rightarrow m_1^2$.

and m_0^6, m_0^7, m_0^8 are pending. B_0^{AA} is deliverable since it contains eight pending messages. The maximal initial segment of the set of pending messages is the set of all pending messages. Hence

$$deliver(B_0^{AA}) = \langle m_0^1, m_0^2, \ldots, m_0^8 \rangle.$$

$B_1^{AA} = mark\_delivered(B_0^{AA})$ is obtained by marking these messages as delivered and forming $D_1 = D^{B_1^{AA}}$, as illustrated in Figure 11.57 where the messages in D_1 are marked with the number 1. The pending messages of B_1^{AA} are marked with p. These pending messages, together with D_1 do not form an initial segment, for example because $m_2^4 \longrightarrow m_1^7$. The maximal set of pending message that form with D_1 an initial segment is

$$m_1^3, m_1^4, m_1^5, m_1^8, m_1^6,$$

and these messages, in that order, form $deliver(B_1^{AA})$. These messages are marked with 2 in Figure 11.57. Form now $B_2^{AA} = mark\_delivered(B_1^{AA})$. The set of delivered messages D_2 of B_2^{AA} comprises those messages marked with 1 and 2. The first-undelivered messages of B_2^{AA} are $m_1^1, m_1^2, m_2^3, m_2^4, m_2^5, m_2^6, m_1^7, m_2^8$. These are the pending messages as well, since all messages are registered. The maximal subset forming with D_2 an initial segment is

$$deliver(B_2^{AA}) = \langle m_2^4, m_1^7, m_2^3, m_2^5 \rangle.$$

These messages are marked with 3 in Figure 11.57. Only in the fourth round are m_2^1, m_1^2 and the remaining messages delivered.

Consider now process P_3 (for example) to see how it delivers its messages. Up to moment t_5, it did not receive neither m_0^1 nor m_0^2, and hence it cannot make the first delivery operation before t_6. At t_6 it delivers the first round (marked 1), but it has to wait until t_8 for the second round. Let's consider now the Early Delivery to see how it improves on the All-Ack Protocol.

Again we consider the fully registered *Data_Buffer* of Figure 11.56, but now it is called $B_0 = B_0^{ED}$. We want to determine the delivery ordering obtained from the derived sequence. First, $F_n = fish(B_0)$ is formed by picking the first-undelivered message in each process. Since no process is delivered in B_0, $m^p = m_0^p$ is the first undelivered message in P_p, and as B_0 is fully registered, F_0 is a full FISH.

$$F_0 = (Id, R, \longrightarrow)$$

where $R = Id$ and $m \longrightarrow n$ iff $m \in \{3,4,5\}$ and $n \in \{6,7,8\}$. The candidates of F_0 are $\{1,\ldots,5\}$, and $\{6,7,8\}$ are the secondary processes. Thus $\#\bar{c} = 3$, and (for $\alpha = 0.5$) $\Phi_{F_0} = 1.5$. Since $V(1) = V(2) = \emptyset$, 1 and 2 are excluded. Since $V(p) = \{6,7,8\}$ for $p \in \{3,4,5\}$, $|V(p)| = 3 > 1.5$ and hence 3, 4, 5 are sources. F_0 is deliverable (being full), and the deliverable processes are the sources. by definition, a data buffer is deliverable iff its fish is, and hence B_0 is deliverable. Thus

$$deliver(B_0^{ED}) = \langle m_0^3, m_0^4, m_0^5 \rangle,$$

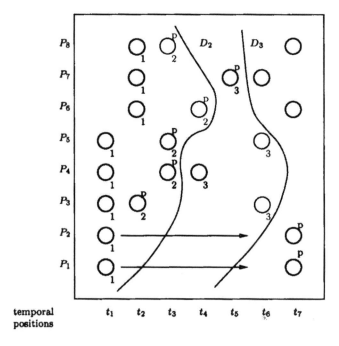

FIGURE 11.57. Deliveries in the All-Ack Protocol. Messages in $D_1 = D^{B_1}$ are marked 1. Pending messages are marked p, messages in $deliver(B_1)$ are marked 2. Messages in $deliver(B_2)$ are marked 3. m_1^2 is not in this set since $m_2^7 \to m_1^2$.

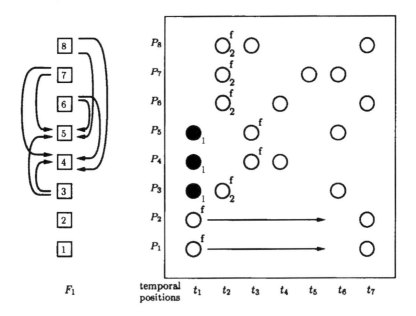

FIGURE 11.58. B_1^{ED} and $F_1 = fish(B_1)$. The messages that are already delivered in B_1 are marked 1. The candidates of F_1 are $\{1, 2, 3, 6, 7, 8\}$. So $\Phi_{F_1} = 1$, and the sources are $\{3, 6, 7, 8\}$. The messages in $deliver(B_1)$ are marked 2.

and $B_1^{ED} = mark\_delivered(B_0)$ is obtained by defining $D_1 = \{m_0^3, m_0^4, m_0^5\}$ as delivered. In Figure 11.58 these delivered messages are marked 1, and the first-undelivered messages are marked with f. The FISH structure $F_1 = fish(B_1^{ED})$ is portrayed in Figure 11.58 as well, and we see that

$$deliver(B_1^{ED}) = \langle m_1^3, m_0^6, m_0^7, m_0^8 \rangle.$$

The messages delivered by B_1^{ED} are marked 2 in Figure 11.58.

$Data\_Buffer$ $B_2^{ED} = mark\_delivered(B_1^{ED})$ is depicted in Figure 11.59. The delivered messages are those marked 1 and 2. The first-undelivered messages are marked f.

$$deliver(B_2^{ED}) = \langle m_1^4, m_1^5, m_1^8 \rangle.$$

Now form $B_3^{ED} = mark\_delivered(B_2^{ED})$ as shown in Figure 11.60. The deliverable message of B_3 are marked 4.

$$deliver(B_3^{ED}) = \langle m_0^1, m_0^2, m_2^4, m_1^6 \rangle.$$

To compare the performance of the two protocols in this scenario, consider a receiver, located for example at P_3, at the different temporal positions. Table 11.3 shows when each message arrives to P_3. For example, we see that at t_3 P_3 has received $m_0^3, \ldots, m_0^8$ and m_1^3. We also see that m_2^3 is only received at t_7. The local $Data\_Buffer$ of P_3 at t_k is denoted B_{t_k}.

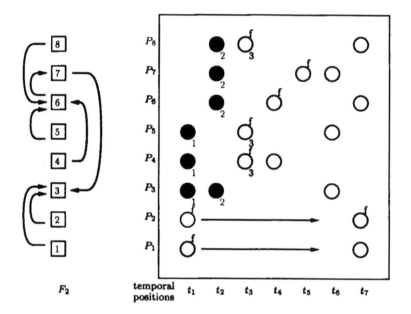

FIGURE 11.59. B_2 and $F_2 = fish(B_2)$. The delivered messages are marked 1 and 2, and the first-undelivered are marked f. This is the first time that some first-undelivered message follows m_0^1, m_0^2 (namely m_2^3). Not all arrows are shown in F_2, and the missing are obtained by transitivity. Thus, for example, $|V(5)| = 3$. $\#c = 5$, $\Phi = 1.5$. The sources are $\{4, 5, 8\}$. 1 and 2 are excluded processes. The messages delivered by B_2 are marked 3.

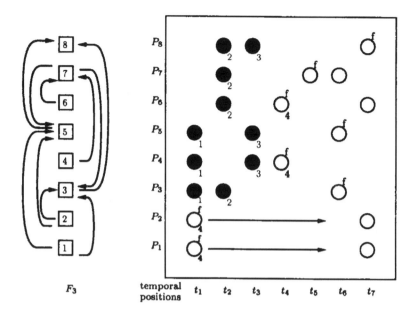

FIGURE 11.60. B_3 and $F_3 = fish(B_3)$. Delivered messages of B_3 are marked 1, 2, 3. First undelivered are marked f. Again not all arrow are shown in F_3 and transitivity should be applied. $V(1) = \{3, 5, 8\}$ for example. $\{1, 2, 4, 6\}$ are candidates. $|\Phi_{F_3}| = 2$. Sources are 1, 2, 4, 6. So $m_0^1, m_0^2, m_2^4, m_1^6$ are delivered.

time	messages received	All-Ack	Early Delivery
t_1	none	none	none
t_2	m_0^3, m_0^4, m_0^5	none	none
t_3	$m_1^3, m_0^6, m_0^7, m_0^8$	none	m_0^3, m_0^4, m_0^5
t_4	m_1^4, m_1^5, m_1^8	none	$m_1^3, m_0^6, m_0^7, m_0^8$
t_5	m_2^4, m_1^6	none	none
t_6	m_1^7, m_0^1, m_0^2	$m_0^1, \ldots, m_0^8$	m_1^4, m_1^5, m_1^8
t_7	m_2^3, m_2^5, m_2^7	none	$m_0^1, m_0^2, m_2^4, m_1^6$
t_8	$m_1^3, m_1^4, m_1^5, m_1^8, m_1^6$	$m_1^3, m_1^4, m_1^5, m_1^8, m_1^6$	m_1^7
t_8		$m_2^4, m_1^7, m_2^3, m_2^5$	m_2^3, m_2^5, m_2^7

TABLE 11.3. The first two column show when each message reaches P_3. The third and fourth column show the deliveries at each temporal position in the All-Ack and Early Delivery protocols.

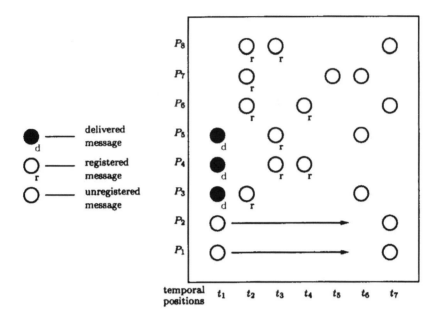

FIGURE 11.61. Local *Data_Buffer* B_{t_4} at t_4. Delivered messages are marked d, registered messages are marked r. $deliver(B_{t_4}) = \langle m_1^3, m_0^6, m_0^7, m_0^8 \rangle$.

B_{t_1} is an empty buffer. B_{t_2} has only three registered messages m_0^3, m_0^4, m_0^5. It is undeliverable since all processes are excluded in its (non full) fish.

B_{t_3} contains no registered messages from P_1 and P_2. The resulting FISH of this buffer is the structure

$$F_{t_3} = (Id, R, \longrightarrow)$$

where $R = \{3, \dots, 8\}$ contains six processes, and $a \longrightarrow b$ iff $a \in \{3, 4, 5\}$ and $b \in \{6, 7, 8\}$. Indices 3, 4, 5 are candidates (minimal registered) and thus $\#c = 3$. Hence process $\Phi = 5 \times 0.5 = 2.5$. So $\#\bar{r} = 2 < \Phi$, and as each candidate is a source, F_{t_3} is deliverable. Hence P_3 delivers m_0^3, m_0^4, m_0^5 at t_3. This is shown in the fourth column in the table.

The local *Data_Buffer* B_{t_4} at time t_4 is described in Figure 11.61. Let $F_{t_4} = (Id, R, \longrightarrow)$ be the corresponding FISH. Then $R = \{3, 4, 5, 6, 7, 8\}$ is the set of registered processes, and $a \longrightarrow b$ holds iff $a \in \{3, 6, 7, 8\}$ and $b \in \{4, 5\}$. Thus $\#\bar{c} = 4$, $\Phi = 2$, $\#\bar{r} = 2$ and F_{t_4} is deliverable with 3, 6, 7, 8 as deliverable processes. Thus P_3 delivers m_1^3, m_0^6, m_0^7, m_0^8 at t_4.

At t_5 two additional messages arrive to P_3, m_2^4 and m_1^6. The resulting *Data_Buffer* is in Figure 11.62. The corresponding FISH has four registered processes: 4, 5, 6, 8. Candidates are 4, 5, 8, and $\Phi = 2.5$. Condition $\#\bar{r} \leq \Phi$ does not hold and this data buffer is not deliverable.

At t_6 three new messages arrive: m_0^1, m_0^2, and m_1^7. The Early Delivery Protocol at t_6 delivers m_1^4, m_1^5, m_1^8. Though m_0^1, m_0^2 are candidates, they are excluded and thence not delivered. (See Figure 11.63). At t_7 the situation

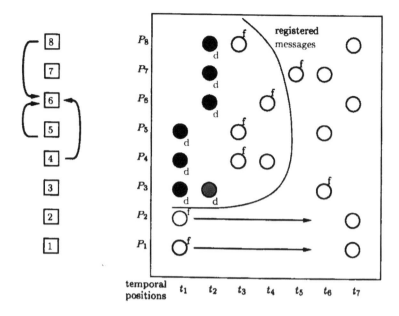

FIGURE 11.62. Local *Data_Buffer* of P_3 at t_5 and the corresponding fish. Registered messages are shown encircled. B_{t_5} is not deliverable.

changes.

At t_7 three new messages arrive to P_3: m_2^3, m_2^5, m_2^7. The resulting *Data_Buffer* and fish is in Figure 11.64. $\#\bar{r} = 1$. $\#c = 4$. $\Phi = 2$. All candidates are deliverable, and hence $m_0^1, m_0^2, m_2^4, m_1^6$ are delivered.

Four new messages arrive to P_3 at t_8: m_1^1, m_1^2, m_2^6, m_2^8. All processes excepting P_4 are registered in the resulting FISH. So $\#\bar{r} = 1$. There is a single candidate, namely 7, and so $\#\bar{c} = 7$ and $\Phi = 3.5$. Condition $\bar{r} < \Phi$ holds, and $V(7) = \{1, 2, 3, 5, 6, 8\}$ shows that 7 is a source and hence deliverable. It is interesting to note that the next delivery cycle is done still at t_8 with no need for more messages. The resulting FISH, after m_1^7 has been delivered, is denoted F. Process 4 is still unregistered, and $a \longrightarrow b$ holds iff $a \in \{3, 5, 7\}$ and $b \in \{1, 2, 6, 8\}$. So $\#\bar{c} = 5$, and $\Phi = 2.5$. Hence m_2^3, m_2^5, m_2^7 are deliverable.

Of course, this example was chosen to illustrate the benefit of the Early Delivery Protocol, and the reader can find examples where it is the All-Ack Protocol that works much better. Dolev, Kramer, and Malki [14] have experimental results (and some rough calculations) which indicate that their protocol is an improvement over the All-Ack Protocol. Since our protocol is somewhat different, it is not clear whether and to what extent these results carry over. Experimental results are very important for the engineering of networks, but theoretical understanding, because of its flexibility and greater range of applicability, is invaluable. I believe that a probability theory of concurrency and interprocess communication has yet to be formed. Such a theory will formulate and calculate (for example) the effects of changing parameter α in the Early Delivery Protocol,

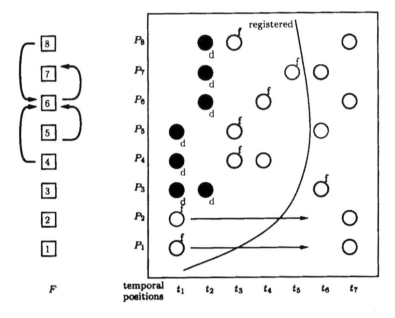

FIGURE 11.63. Local *Data_Buffer* of P_3 at t_6, and $F = fish(B_{t_6})$. The registered processes in F are $\{1, 2, 4, 5, 6, 7, 8\}$. Candidates are $\{1, 2, 4, 5, 8\}$. Thus $\Phi = 1.5$, $\#\bar{r} = 1$ and F is deliverable: 1, 2 are excluded, and 4, 5, 8 are sources. Thus 4, 5, 8 are deliverable.

and will teach us about the differences between these protocols.

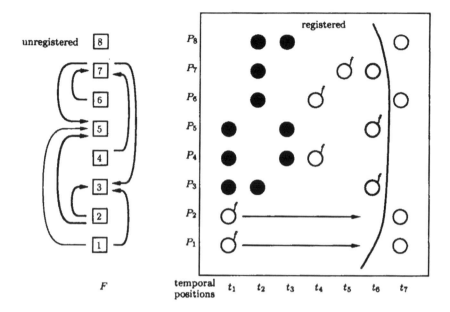

FIGURE 11.64. At t_7 the resulting fish F contains four registered messages.

$M\_SendEvent(S)$ 207	$v = ts(m_i^p)$ 207
$M\_DeliverEvent(D)$ 207	Empty data buffer 210
Fully registered 210	pending 212
candidate 212	*deliverable(B)* 213
derived sequence 213	global buffer 214
E^B 216	evidence set 216
fish(B) 216	registered 216
candidate 216	secondary 216
#$\bar{r}$ 216	#c 216
source 217	excluded 217
deliverable fish and process 217	early delivery 217
all-ack 217	end-extension 217
deliverable data-buffer 221	

TABLE 11.4. Chapter's index.

Epilogue: Formal and informal correctness proofs

I would like to conclude with a note about the origin and motivation for this work. When I turned towards computer science my co-workers were Menachem Magidor and Shai Ben-David, and the first questions we studied were about mutual exclusion protocols, and issues envisaged by L. Lamport concerning the concept of global time. Most of our results are not suitable for an introductory text, but their influence is evident, and the place of system executions in my work can be traced back to that period and is a testimony to the influence of Lamport's [20] on my work. At that time I was interested in atomic register problems, and it was expressly in order to prove the correctness of the protocol found in [1] that the usage of higher-level properties developed.

The quest for a mathematical framework within which properties of protocols could be proven has been a central motivation for the work presented here. To describe this motivation I would like to quote from a work written with Shai Ben-David.[1] Usually it is not good manners to quote from your own papers, but this one was never published, so that I feel like honoring a neglected friend.

> Proofs in mathematics can have different aspects, but two are prominent. The first one, of course, is that a proof is used to validate a statement—to establish a theorem. Yet there might be another aspect of a proof: a proof is a way to convey ideas and to *explain* why things work. So, for example, when we see a proof of Pythagoras's Theorem, not only are we convinced that the theorem holds, but we also learn something else about triangles; we have a better understanding of right-angled triangles. Usually, there is no sharp distinction between these two aspects and they are blended together—sometimes with more emphasis on the one than on the other. So when a mathematician wants to explain the idea of a proof he or she does not present a formal proof, but outline a description of one, or of what can be transformed into such. Usually,

---

[1] *Informal and Formal Correctness Proofs for Programs (for the critical section problem)*, November 1987.

arguments are given in order to explain and to communicate, and sometimes examples are given or pictures. Then, if more details are needed, they are provided; until, at least in theory, a complete formal proof can be given—one which is the counterpart of the informal one.

Unfortunately, the situation does not seem to be like that in computer science. The proofs which we have seen for showing the correctness of parallel algorithms are hard to follow, and very often even after the sequence of statements which constitute the proof has been followed line by line, the algorithm remains as enigmatic as before.

We feel that there is a need for a way to present intuitive and informal proofs for program correctness, and in a form which can easily be altered to a formal proof if need arises. The framework for correctness proofs should be large and flexible enough to allow both informal and formal proofs.

We did not have an answer to this need at that time, and this book represents my effort to find such a framework. Certainly other researchers have had similar motivations, and solutions were offered by several prominent researchers. Let me quote for example from the book by K. M. Chandy and J. Misra ([12] page xi).

> One can reason about the unfolding computations of a program (this is called operational reasoning) or one can focus, as much as possible, on static aspects (such as invariants) of the programs. We favor the static view here for three reasons. First, we made more mistakes when we used operational reasoning. ... Second, we have found it hard to convince skeptics about the correctness of our programs by using operational arguments. ... Third, our operational arguments tend to be longer.
>
> Operational reasoning has value. Again, being very subjective, we have found that the flash of insight that sparks the creation of an algorithm is often based on operational, and even anthropomorphic, reasoning. Operational reasoning by itself, however, has gotten us into trouble often enough that we are afraid of relying on it exclusively. Therefore we reason formally about properties of a program, using predicates about *all* states that may occur during execution of the program.

Chandy and Misra develop a logic for reasoning about sequences of program states, and our monograph can be seen as an attempt to secure a formal foundation for operation reasoning.

References

[1] U. Abraham, On interprocess communication and the implementation of multi-writer atomic registers, *Theoretical Computer Science* 149 (1995) 257-298.

[2] U. Abraham, On system executions and states, *J. of Applied Intelligence* 3 (1993) 7-30.

[3] U. Abraham, What is a state of a system (an outline), in : M. Droste and Y. Gurevich eds, *Semantics of Programming Languages and Model Theory*, Vol. 5 of *Algebra, Logic and Applications* (Gordon and Breach, 1993) 213-243.

[4] U. Abraham, Bakery algorithms, in: H-D Burkhard, L. Czaja and P. Starke eds., Proceedings of the Concurrency, Specification and Programming Workshop, Nieborow 1993 (Wydawnictwa Uniwersytetu Warszawskiego, Warszawa, 1994) 7-40.

[5] U. Abraham, S. Ben-David, and M. Magidor, On global-time and inter-process communication, in: M. Z. Kwiatkowska et al., eds, *BCS-FACS Workshop on Semantics for Concurrency*, Leicester (1990) 311-323 (*Workshops in Computing*, a Springer-Verlag Series edited by C.J. van Rijsbergen).

[6] U. Abraham, S. Ben-David, and S. Moran, On the limitation of the Global-Time assumption in distributed systems (extended abstract), in S. Toueg et al., (eds.) *Distributed Algorithms*, WDAG '91 Proceedings, Lecture Notes in Computer Science 579, pp 1-8, Springer, 1991.

[7] B. Alpern, and F. B. Schneider, Defining liveness, *Information Processing Letters*, 24:4, pp.181-185, 1985.

[8] F. Anger, On Lamport's interprocess communication, *ACM Transactions on Programming Languages and Systems* 11 (1989) 404 - 417.

[9] S. Ben-David, The global-time assumption and semantics for distributed systems, in: *Proc. 7th Ann. ACM Symp. on Principles of Distributed Computing* (1988) 223 - 232.

[10] K. Birman, and T. Joseph, Reliable Communication in the Presence of Failures. *ACM Trans. Comp. Syst.*, 5(1):47-76, February 1987.

[11] B. Bloom, Constructing two-writer atomic registers. 6th ACM Symp. on Principles of Distributed Computing, pp. 249 - 259, 1987.

[12] K. M. Chandy, and J. Misra, *Parallel program design : a foundation*, Addison-Wesley, 1988.

[13] E. W. Dijkstra, Co-operating Sequential Processes, in F. Genuys (ed) *Programming Languages*, Academic Press, London, 43—112, 1968.

[14] D. Dolev, S. Kramer, and D. Malki, Early Delivery Totally Ordered Multicast in Asynchronous Environments *(preprint from the Hebrew University, 1992)* and in *23nd Annual International Symposium on Fault-Tolerant Computing (FTCS)*, Toulouse, France (1993) 544-553.

[15] S. Dolev, A. Israeli, and S. Moran, Self Stabilization of Dynamic Systems Assuming Only Read Write Atomicity, *Distributed Computing*, Vol. 7, pp. 3-16, 1993.

234

[16] P. C. Fishburn, *Interval orders and interval graphs*, Wiley, New York, 1986 (Wiley-Interscience series in discrete mathematics).

[17] V. K. Garg, and A. I. Tomlinson, Causality for Time: How to Specify and Verify Distributed Algorithms, *IEEE Symposium on Parallel and Distributed Processing*, Dallas, Texas, October 1994.

[18] S. Kramer, *Total ordering of messages in multicast communication systems*. Master's thesis, Hebrew University, Jerusalem, 1992.

[19] L. Lamport, Time, clocks, and the ordering of events in a distributed system, *C. ACM*, **21**:7 (1978) 558-565.

[20] L. Lamport, On interprocess communication, Part I: Basic formalism; Part II: Algorithms, *Distributed Computing* **1** (1986) 77 - 101.

[21] L. Lamport, and F. B. Schneider, (Chapter 5) In: *Distributed systems*. M. Paul and H. J. Siegert (eds) Lecture Notes in Computer Science 190, Springer 1985.

[22] Z. Manna, and A. Pnueli, *The Temporal Logic of Reactive and Concurrent Systems*, Vol.1 "Specification" Springer Verlag, 1992.

[23] . M. Melliar-Smith, L. E. Moser, and V. Agrawala, Broadcast Protocols for Distributed Systems. *IEEE Trans. Parallel & Distributed Syst.*, (1), Jan 1990.

[24] S. Owicki, and D. Gries, An axiomatic proof technique for parallel programs I, *Acta Informatica*, Vol. 6, pp. 319–340, 1976.

[25] G. L. Peterson, Myths about the mutual exclusion problem, *Information Processing Letters*, vol. 12, pp. 115–116, 1981.

[26] G. L. Peterson, and M. J. Fischer, Economical Solutions for the Critical-Section Problem in a Distributed System, Proc. of Ninth ACM Symposium on Theory of Computing, 1977, pp. 91-97.

[27] V. Pratt, Modeling Concurrency with Partial Orders, *Internat. J. of Parallel Programming* **15** (1986) 33 - 71.

[28] A. Silberschatz, J. L. Peterson, and P. B. Galvin, *Operating System Concepts*, third edition, Addison-Wesley, 1991.

[29] M. Raynal, and M Singhal, capturing causality in distributed systems, *Computer*, vol. 29, No. 2, pp. 49—56, 1996.

[30] A. S. Tanenbaum, *Computer Networks*, Prentice Hall, Englewood Cliffs, NJ, 1989.

[31] N. Wiener, A contribution to the theory of relative position, *Proc. Camb. Philos. Soc.* **17** (1914) 441-449.

[32] G. Winskel, An introduction to event structures, in *Linear Time, Branching Time and Partial Order in Logics and Models for Concurrency*, edited by J. W. de Bakker et al. Lecture Notes in Computer Science, **345**, pp. 364-397, Springer.

Index

$\models$ models 11
$S|L$ reduct 13
Left_End Right_End 25
$\prec_L$ 24
$\vdash_L$ 25
$\parallel$ 32, 58
$\prec$ 33
ω 35
locations(X) 52
initial(X) 55
    states(X) 55
$Tr(\alpha)$ 55
$Tr(X)$ 57
$\mathcal{H}(X)$ 58
$<_{steps}$ 58
$\Sigma(X)$ 59
$\lceil j \rceil_N$
γ 149, 190
$\xrightarrow{causal}$ 194
$\xrightarrow{D}$ 197

A

antichain 26
arrow (of flowchart) 52, 82
assertional semantics 69, 77
assignment 10, 55
attentiveness
    ot a channel 149
    of a net 190
Axioms:
    channel 152
    global time 29
    Data_Buffer
    Delivery 195
    Message Domain 191
      with deliveries 194
    Network 190, 193
    Send/Receive 190
    Sliding Window 157
    TimeStamp Vector 200
    Throw/Catch 40

B

Backus-Naur form of instructions 51
broadcasting 189
buffer cell 114
    circular 133
    conflict free 114
    safe 115

C

causality (preservation at protocol and user
    layer) 194-198
channel (specification, attentive, no dupli-
    cation, non-lossy, order preserv-
    ing, capacity of, orderly operat-
    ing, FIFO, raw) 149-153
communication device 35
concurrent protocol 52
conflict free (buffer) 114

D

Data_Buffer 207
device 39
Dijkstra's semaphore 115
    protocol 126

E

equiped with pairs 7
event
    higher and lower-level 32
    lost send 151
    event structure 28
    semaphore 118
    terminating/nonterminating 27
execution (of a history) 59
expansion of a structure 13
external operation 50, 59
external semantics 48

F

FIFO 151, 152, 187, 198
finite predecessor property 27
first order formula 9
flattening 13
flowchart 52, 69
    combined 54
    higher level/explicit form 80-81, 90
    history and semantics of 57, 81

G

global-time 30-31
    model 27

H

history (of flowcharts) 58, 81
history model 19, 58–60, 66, 96
    induced 82

I

induction "on right-end points" 29
instructions (BNF) 51
interleaving approach 77
interpretation 4
    multi-sorted 5
interval ordering 24-25

L

Lamport's finiteness condition 27
    global-time models 27, 30
    structure 31
    granularity of systems 78
language
    first-order 9
    protocol 49
location variable 55

M

Manna and Pnueli's semaphore 116
Message Domain with Delivery 196
    signature and axioms 192, 193
model 12
    history model 57
    two level history model 97
Mutual Exclusion Property (MUTEX) 20

N

network (non lossy, attentive, FIFO) 191
Network Signature (and axioms) 190
no-duplication (channel) 149
non interference 52

O

operation
    external 37, 48, 57
    terminating 37
orderly operating channel (FIFO) 151

P

pair 7
parameter 38, 50
Pitcher/Catcher 40, 79
procedure (declaration) 90
process 34-36, 66, 118, 190
protocol, concurrent 50, 89
    language 50
Protocols:
    Aimless 100
    Dijkstra (semaphore) 126
    KanGaroo and LoGaroo 92
    Multiple Process Mutual Exclusion 129
    Multiple Producers 144, 160
    Peterson's Mutual Exclusion 16
        revisited 69
    Peterson–Fischer mutual exclusion 107
    Pitcher/Catcher 60
        revisited 79
    Producer/Consumer (generic) 112
    Redressing 154, 155
    Sliding Window 170
    Unbounded *ENQUEUE/DEQUEUE* 133
        Bounded 142
    uniform delivery 198, 202
        All-Ack
        Early Delivery (Dolev, Kramer, Malki)
        99

R

raw channel 149
reduct 13
register
    local 106
    regular 36
    safe 36
    serial 35
    specification 35
representation of partial orderings 25
    of global-time models 28
return function (ω) 35
Russell–Wiener property 25-26

S

safeness (of buffer) 115
satisfaction 8
semantics 48, 58-59
    assertional 69
    of flowcharts 55
semaphore 115
    P/V theory of 122
    synchronization with 126
sequence over A
signature 3
    multi-sorted 4
sort
specification
    buffer cell 114
    channel 152
    device 36
    ENQUEUE/DEQUEUE 114
    register 34
    semaphore (textbook) 119
state regular/unreachable, initial, global 54
    induced 82
    of higher-level flowcharts 80
    over X 55
step 58
    substep 83
structure 4
    background 38, 49

    composite 13
    equiped with pairs 7
    isomorphism of 5
    state and transition
    substructure 12
system 30
system execution 30, 33
    two-level 90

T
term 9
terminating operation 39
theory 11
throw/catch operations 40
time (Achilles) 41
time-stamp 199

transition terminating/non-terminating 56
type

U
Uniform Delivery 196
unrestricted semantics 48

V
value parameter
variable 9, 49, 79, 133

W
Wiener 26
Winskel's event structures 30

Z
Zeno's theorem 46

Printed and bound by CPI Group (UK) Ltd, Croydon, CR0 4YY

23/10/2024

01777667-0011